# FROM MY PAST LIFE

## MOSTLY FLUTE MUSIC AND MATH WRITING

**Franz Rothe**

978-1-965552-39-1 (Paperback)
978-1-965552-40-7 (E-book)

BOOKWRIGHTS
HOUSE

admin@bookwrightshouse.com
☎ (213) 286 6700

# Preface

HOW DID THE DECISION to write such a book come about?

A smart schoolmate invited each of our final class members to write about themselves. Using some short essays I enjoyed writing, I had gathered enough material for such a booklet. The task of creating a comprehensive autobiography seemed too challenging. So I decided to gather my short observations, put them into a loose context and make a book, even though it is clearly incomplete.

Why do you think this book will appeal to more people than only your closest friends?

The description of professional growth, and the music playing which I have practiced with endurance during my entire life can motivate and benefit a broader audience.

Is it easy or difficult for the reader to tell which parts of the stories are invented and which are true?

I think that is easy enough to ensure that it does not become cheating. To make the entire book more personal, I have included imagined parts.

Charlotte, February 2025

Keywords: school memories, literary criticism, humor, cat love, philosophy of mathematics, flute music, Bach, Mozart, Kafka.

Genre: biography, flute literature, mathematics literature.

# *About the Author*

━━━━━━━━━━━━━━━━━━━━━━━━━━━━━━━━  ● ● ●

FRANZ ROTHE GRADUATED FROM high school in Karlsruhe and studied mathematics, physics and music there. Graduated with a diploma in mathematics from the E T H Zürich, a doctorate in Tübingen. After some changes in life, a professorship at the University of North Carolina at Charlotte, USA. In addition, Rothe and pianist Thomas Turner have developed a repertoire of classical music for flute and piano, and have recorded and released three CDs. This collection also contains several of their own transcriptions. Now Rothe is retired and keeps writing books about mathematics, and too, just for entertainment.

The audio book version starts with a nice song, a present by Alvi White.

Recalling Past Life

# Contents

# Chapter 1:
# A Teacher in Good Mood

● ● ●

## 1.1  Thales' Theorem

IF YOUR BOOK WERE adapted into a movie, how would you envision its opening scene to captivate audiences?

The opening scene shows the young professor at the blackboard proving Thales' Theorem about the right angle in a semi-circle.

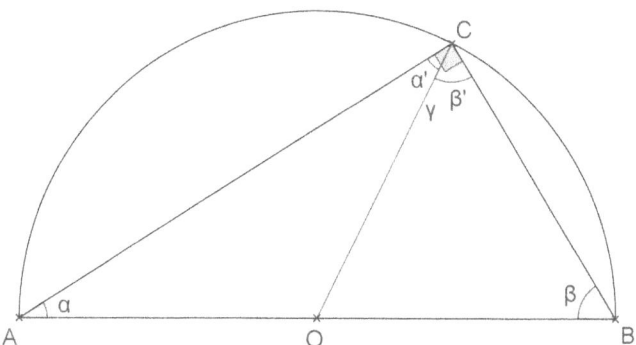

**Figure 1.1.1:** Thales Theorem.

**Theorem 1 (Thales' Theorem).** *"The angle in a semicircle is a right angle." More precisely stated: If an angle has its vertex C on a circle, and its sides cut the circle at the two endpoints A and B of a diameter, then angle ACB with* *the vertex C lying on the semicircle, and its sides passing through A and B, is a right angle.*

*Proof of Thales' Theorem in modern manners.* Draw the triangle *ABC* and, as an extra for the proof, the line from the center *O* to the vertex *C*. The triangle is partitioned into two triangles. Since the segments *OA*, *OB* and *OC* from the center of the circle to the vertices of the triangle are congruent, we get two isosceles triangles. The base angles of an isosceles triangle are equal by Euclid I.5 . In modern parlance, we say:

The base angles of an isosceles triangle are congruent.

We use that fact at first for the triangle *AOC*. Hence

$$\alpha = \angle OAC \cong \angle OCA$$

Secondly, we use Euclid I.5 for the triangle *COB*. Hence

$$\beta = \angle OBC \cong \angle OCB$$

By angle addition at vertex *C*

(1) $$\gamma = \angle ACB = \alpha + \beta$$

Next we use Euclid I.32, which tells us:

The sum of the angles in a triangle is two right angles.

Because $\alpha$, $\beta$, $\gamma$ are just the angles in $\angle ABC$, we conclude that

(2) $$\alpha + \beta + \gamma = 2R$$

I have used the letter *R* to denote a right angle. Using formulas and (2) together yields $\gamma + \gamma = \alpha + \beta + \gamma = 2R$, and hence $\gamma$ is a right angle, as to be shown.

Thales's lived about 624-546 B.C., in Miletus, a Greek island along the coast of Asia Minor. These dates are known rather precisely, because, as reported by Herodotus, he predicted a solar eclipse, which has been determined by modern methods to have occurred on May 28th of 585 B.C.

Tradition names Thales of Miletus as the first Greek philosopher, mathematician and scientist. He is known for his theoretical as well as practical understanding of geometry. Most important, he is credited with introducing the concept of logical proof for abstract propositions.

Traditions surrounds him with legends. Herodotos mentioned him as having predicted a solar eclipse, which put an end to fighting between the Lydians and the Medes.

Aristotle tells this story about him:

Once Thales somehow deduced that there would be a great harvest of olives in the coming year; so, having a little money, he gave deposits for the use of all the olive presses in Chios and Miletus, which he hired at a low price because no one bid against him. When the harvest-time came, and many were wanted all at once and of a sudden, he let them out at any rate he pleased, and made a quantity of money.

Plutarch tells the following story:

> Solon who visited Thales asked him the reason which kept him single. Thales answered that he did not like the idea of having to worry about children. Nevertheless, several years later Thales anxious for family adopted his nephew Cybisthus. Thales went to Egypt and studied with the priests. While he was in Egypt, he was able to determine the height of a pyramid by measuring the length of its shadow at the moment when the length of his own shadow was equal to his height. Thales is said to have proved some simple theorems of geometry, as well as the not so obvious theorem about the right angle in a semi-circle. As stressed by David Park in his nice book *The Grand Contraption— the world in myth, number, and chance* the story raises an important point, whether or not Thales really invented the proof:
> 
> Babylonians and Egyptians had a number of mathematical tricks. For example, Babylonians knew this proposition, as well as the Pythagorean theorem

a thousand years before Thales and Pythagoras found them. If they were known, they must have been proved, but there is no sign that anyone thought the proofs were important enough to preserve. Whoever set the process of proof at the center of the stage is the founder of all the mathematics since then, and if it is not Thales it was someone who lived not long afterwards.

## 1.2    Plato and Slavery

Can you describe the world or setting of your book and how it could translate visually on screen?

A young assistant professor, obviouly myself lecturing, again in the second scene, now getting distracted to make a digression about Plato's travel to Sicily.

The professor has been lecturing for half an hour. Now both blackboards are fully covered with his notes. He picks up the eraser and hesitates.

"In ancient Athens, there would have been a slave for wiping the blackboard." It slipped out of him like that.

By the way, do you know that Plato sailed to Sicily three times, with the intention of founding an ideal state there? The third time, he meets the tyrant Dionysus. Some historians speculate he had promised to come back. Now he has just explained his plan to the tyrant.

"So you tell me that in your state, every man and woman does the work for which they are best suited."

"Yes, Mr. Dionysos."

"Let's see." The tyrant approaches Plato and grabs him a bit at the upper arm. "Healthy guy, strong muscles."

He explains: "It is really convenient that my large slave market will open just tomorrow morning. Water carriers are urgently needed for my thriving economy. We will offer you immediately on the right side at the entrance gate. Excellent sales prospects! And before the sun is high up at noon, you no longer need to make push-ups. Then you can do something more useful."

And whether Plato's friends had listened right behind the door and were able to free him on the spot, or whether they only bought Plato free on the slave market the next morning, this is not known. In any case, Plato and his friends managed to sail back to Athens. And only after his attempt to establish an ideal state had failed for a third time, did he found the academy in Athens. It developed to the first university.

"And did he continue to advocate slavery afterwards?" asks a student in the back row.

"So little has been handed down safely from ancients times. But I think, yes." The professor answers and begins to wipe the blackboards.

## 1.3    Remembering Karl-Peter Hadeler

Professor Dr. Karl-Peter Hadeler had been my doctoral supervisor, and two years later for many years my employer at the University of Tübingen. Because of the overwhelming expectations and the strangeness I experienced at the E T H, and the imminent death of my father, I had not been quite happy during the past time. I was new at in Tübingen. Always since the start of my dissertation, Hadeler has been an always reachable contact person for me. For example, Hadeler had translated for me important publications by Kolmogoroff et.al. from Russian and had spoken them on tape. Soon I got to appreciate the hospitality of Helgard Hadeler and her family. Thanks to all their friendliness, I felt much more at home than I had in Zürich during the years before. Hadeler's chair for biomathematics had a circle of long-term and short-term employees. The seminar for biomathematics brought many contacts and constant motivations for my work. Despite this diversity, Hadeler has always encouraged me, as well as the other coworkers, above all to publish independently, starting in small documented steps. Hadeler himself contributed regularly to the work of the assistants.

Over time, it became clear to me that it was the tasks of two professorships that Prof. Hadeler wanted to fulfill and could achieve. In addition to the professorship for mathematics, especially in the Russian school, a second one for mathematical biology, especially in its current developments. Only a first-class man with his exceptional intelligence, perseverance, and human prudence could cope with such enormous demands.

### 1.3.1 Excellent Behavior

From the long years as an assistant at the institute of Professor Karl-Peter Hadeler, also called KPH, I remember only few critical remarks towards to me. Admittedly, more than I want to frankly mention here. He liked to call me "Emperor Franz" and warned me not to exaggerate. A reminder that the scientific career must move forward quickly was often due.

"Everything takes forever with you." The most violent remarks I remember are:

"I am still your employer" and "You are really a strange person."

By the way, I remember an anecdote that KPH once told at the institute tea. At the train station, he had just helped a young lady and carried her suitcase. Where does she want to go? To a congress for teachers of German and German scholars. German was KPH's favorite subject at school alongside mathematics. For whatever reason, he took his time and stayed at the conference for the coffee break. With his good nursery, he has just joined another participant, introducing himself as Dr. Hadeler, and a conversation started. They are talking about the relationships between Goethe and Riemer and between Goethe and Eckermann. Out of genuine interest, KPH contributes some details. His counterpart is pleased with the stimulating conversation and the two gentlemen have a good time with each other. Until the conference participant asks what he is currently working on. KPH replies that he is a graduate mathematician and an assistant in Hamburg. The conference participant hints another greeting with his hand and says:

"Dr. Hadeler, I am very pleased to have met you," gets up and leaves.

Remark. After 1990 I had to leave Tübingen. I came back only for short visits and in my dreams. It should therefore be left to better experts to appreciate Hadeler's later achievements. It is not the place to pinpoint when I realized the scope of his tasks mentioned above—nor how far or not I ultimately have exploited all these options for myself. It should only be mentioned that in April 2017, Springer's book by K.P. Hadeler on cellular machines will be published posthumously. K.P. Hadeler has left us on February third, 2017.

## 1.4 Dreaming of Tübingen

What do we not all experience while dreaming! I guess: when the landlord slumbers fine and sweet, his servants will soon notice. They are right goblins and begin doing all kinds of nonsense. There is turmoil, and since the Lord cannot intervene, it sometimes gets a real turmoil. As for example in my following dream.

Still another trip, tonight. I drag the trolley luggage somewhere through a small old town, is it Tübingen, or maybe L'Aquila? In front of me is a row of pretty old houses, with a church skillfully installed. So L'Aquila? But my car had a breakdown, and can only be checked tomorrow morning and maybe repaired. Where should I stay? Two students come by and animatedly talk to each other, apparently about Gerhard Steiff.

"You have to pronounce say everything very clearly, so he wants it."

I ask if they know of a night's accommodation. For interns only, they do an internship with Steiff.

"How nice," I mean, still embarrassed.

The suitcase is getting heavier. You have to stretch your arms forward as if for the Roman greeting, then the suitcase rolls a bit. I'll try it. But I have to get up this defaced free staircase. Half of it is bricked up with cobblestones. Really Swabian, so pitiful it does look. Finally, I get into a building and a hall. Some people are scattered around. You can look outside through a window into a park with sculptures. The people in here move so slowly as if they would practice something. I see Mr. Dahmen from biocybernetics. It seems like he's trying to look similar to the equestrian statue out there. He looks like Abraham anyway, who in turn almost looks like God. When I ask for accommodation, I get asked back:

"Are you doing an internship with Stalin or Beethoven?"

The blow almost hits me when I consider that. But my cat has jumped on the bed, nothing has happened to me. The animal obviously feels that I should wake up now. No internship, no broken car, comes to my mind immediately. It's time for cat food. Yes, I'm safely back in bed. It remains only a bit of a headache for two days from my bang. In an emergency, the medical expression for this is more of a stroke. May it be so harmless for everyone.

So for explanation: I have not seen Mr. Dahmen for a long time. Most recently, when I had to leave Tübingen and he intended to buy my used car—an Audi. Many years later, on the trip from Heidelberg to Jena, I was again in Tübingen, only for a few hours. KPH, my employer and doctoral supervisor, bought a gift for Tilman while walking through the old town and gave it to me. He bought it in the small shop of Ursula Dahmen in the downtown Tübingen. KPH and I still knew Ulrike as Ulrike an der Heiden. Uwe an der Heiden had been my colleague; I can confirm that he looks optimistic and serious, but not like Abraham. The latter may have been important for Ursula. You don't know that details anyway.

Without much to be interpreted, one can see from the above lines that the Tübingen years were the happiest in my life. The work and relationship with KPH were decisive for this. On top of that, musical activities. I also knew Ulrike and Mr. Dahmen through their joint participation in the university choir under Alexandro Sumski. After a few years,

I turned my back on Sumski. Over time, the constant tension between aspiration and reality was more than annoying and sometimes even ridiculous under Sumski. I was able to join Gerhard Steiff's chamber choir. He became my second caregiver in Tübingen. Miraculously, almost without other external circumstances, all participants felt more balanced under Steiff.

# Chapter 2:
# Back to School

$\bullet$ $\bullet$ $\bullet$

## 2.1   The Invitation

THANKS TO THE DILIGENCE of Reinhold, everybody from our class who wants it may meet in Sulzburg next June. He invites us:

My dear Ones,

I like to hear that—your decision to come to Sulzburg. And Michael will be with us, too. Concerning the accommodations: Rebstock and Dormitorium are centrally located and only 50 m apart from each other (flat route!). Dinner can be taken at the restaurant of the *Rebstock*, they have a separate room, similarly the breakfast. But we may also have breakfast with me at home, no problem. The *Waldhotel* is located three kilometers outside the town, very beautiful, has enough space for all of us, but is not exactly inexpensive. Decide by yourself!

Concerning the program, I have not yet given myself detailed thoughts, because for me at our age more than ever applies Matthew 6,V 34. Here is bible verse quoted:

So don't worry about tomorrow, because tomorrow will take care of itself! Every day has enough with its own plague!

But a tour of our old Ottonian church (more than a thousand years old) and our former synagogue is clearly possible. Moreover, your thematic suggestion, dear Mai, is appropriate and helps to free oneself from indulging in times past.

—Kind regards

Mai is happy to write:

I imagine the program to be similar to last time. Coffee, tour around downtown Sulzburg, dinner with open end. A topic for conversation could be:

"What would I do differently in my life if I came fresh from school again."

That could open up interesting discussions. You see: I pondered about it a little more.

In the audio book you can listen to *Here is Spring* from the "Eight Early Songs" by Claude Debussy, played with flute and piano.

## 2.2 Who Talks so Lovely about Me

One thing is clear to me,
Franzel is a child prodigy!
He can do math, chemistry and physics,
but he never would get Latin tricks.
Also, in English and in French
it doesn't always work in his bench.
And about German, I hardly dare to speak,
there Franzel is quite weak.
But his eyes soon get brighter
playing a flute piece is lighter.
Because he is a musician keen,
as everyone has heard and seen.

—from the pupil's journal

Already in the old pupil's journal, you had flattered my conviction of being a little genius. Thank you for these pats. But everyone has noticed that it didn't always work properly for me, so I don't really need to repeat it. I still cannot offer a blossom pure success story, only something more like a beer newspaper, in the old tradition. After choosing the mathematical sciences for my profession, voluntarily or not, I got around quite a bit. And it didn't work with a permanent position for a long time. Finally, I reached a professorship in the USA, more precisely Charlotte in North Carolina; just two years before all the real geniuses, plagued and hungry, come from Russia and China to the USA and especially North Carolina and look for a job there, too.

But let me tell one after another. After studying in Karlsruhe and Zürich and a rather long assistantship at the university of Tübingen, I became visiting professor in Chapel Hill, North Carolina for a year; then three years I was working at the Bundeswehr University in Munich. Right from the start, I had the pleasure of being made a savior in need by seasoned officers.

This happened because one of the professors had just dropped more than eighty percent of the students in the intermediate diploma in mathematics. Hence, their superiors rightly worried about the future of the young officers. After three trimesters, the boys thanked me for my lecture. And the dean wrote me a thank-you letter, which since then I have been happy to enclose my application documents. But unfortunately none of these tricks helped me to become a professor in Germany. I rejected a time position that I could have gained in Dortmund two weeks before moving to the USA. I have been an associate professor at the University of North Carolina at Charlotte since 1989.

Because of all the many moves, my wife and I have renovated two apartments and three houses and created eight new gardens during the past 15 years. I could sing a long song about the work. My wife and me and my two sons Tilman and Dietrich didn't like the many moves. After four years in Charlotte, my family preferred to return to Germany in the end. There were several reasons. I still hadn't got a permanent position, but my wife's time to leave her job could not be extended any longer. We both wanted to send our children to school in Germany, and were just homesick, to tell the truth. So I have now become a hiker between the USA and Germany, more precisely Charlotte and Celle.

At the university in Charlotte, my colleagues are all quite easygoing. Concerning research, I now have published approximately 30 publications. As far as teaching is concerned, I can honestly say that I like to give lectures. But the lessons are thought to be much more school-like than I actually expected or wanted. The grades are always so terribly important.

But when I say "notes" I prefer to think about music. Already in the pupil's journal, it was so nicely written about me: "But his eyes get brighter, when he even plays a flute piece." Fortunately, nothing has changed for that matter. I regularly play music with a former professor of piano, Thomas Turner. I also practice piano music on my own grand piano. The pieces of music that please me, and are appropriate, become more and more demanding over the years. And my son Dietrich also participates with his cello. I like to accompany him, hopefully carefully and delicately enough.

## 2.3   School Traumata

Some people have dreams like that for a lifetime: they have to pass the exams for the high school diploma once more, again and again. For my part, I was degraded several times, too, when only in a dream. In the beginning, it was still voluntary, for example, to improve my English knowledge,—but only initially. One day, sooner or later, the teacher says,

"Rothe, you're actually quite right here."

Does this happen only to me? Truly not, if one believes the incidents inspired by famous artists and subsequently somewhat adapted and alienated.

Once more, a school bell is ringing.

The new one sits on the bench that the school servant has just carried into the classroom.

"Ton nom?" the French teacher sounds.

He gets up, his hat falls down to the floor. He is close to crying, and can only stammer with fright. When asked about his name for the first and second time, he only produces unidentifiable clams. Only at the third request does he pump up his lungs and tosses out with full strength:

"Charlbovary"

like a "Sharl" with a complicated rat tail, so to speak. Howling roar and laughter. Everyone mimics him.

"Quiet the class, and until tomorrow you will write fifty times: 'Ridiculus sum'".

At this moment the whole class, from the foremost to the rear-most bench, believes that the new one is ridiculous. What should he ever achieve in life? Yes, the reader will get well informed about it in the following world famous pages. His first wife, who had long since passed away, wanted "just a little more love." And she is financially completely ruined, according to the famous story. Is this dreadful end at least a warning to him for the future? Not at all, the knowledgeable reader knows more precisely.

Another school bell is ringing.

But I want to tell you something about my own school time. In this lecture, our old dear math teacher Lina is covering the quadratic functions. She interrogates the entire class:

"And if you now calculate $y$ equal to $x$-square, —one after the other for $x = 1, 2, 3, \ldots$ and so on?"

Reinhold immediately comes up with a brilliant answer. Well, but Lina continues to drill.

"What are the increases between these numbers?"

He hesitates. Fortunately, Nikolaus helps out. But Lina can't let her darling lie on the left so forgotten. So she asks:

"Now Sylla, explain how the sequence of numbers is continuing."

Reinhold is not embarrassed. He knows it and Lina is enraptured. Her enthusiasm rises to the limit. She calls into the class:

"The Sylla has discovered *the law.*"

At this moment the whole class, from the first to the last bench, believes that the Sylla discovered *the law.* He is certain to get a place among the immortals, perhaps even in heaven.

Another school bell is ringing.

The opera on Sunday evening was wonderful. But now it's Monday morning, B. hasn't prepared anything. The damned alarm clock rattles. A terrible day. B. is late. Lina or a related monster is already guarding the school entrance. He has to go around the house but does not escape her pursuit and curses at the back entrance, either.

The first hour passes so so. But now it's Latin's turn. Heberdinger was not as long at the hairdresser's last night as B. was in the theater. He was able to sleep in, and can now translate quite rightly and properly. The crammer gets bored.

"Buddenbrook, you are also welcome. You have practiced your Chopeng enough on the bench. So, Mr. Buddenbrook?" B. rises laboriously and begins to stutter. "Aurea prima est, ... secundum nullum ..."—

"Yes, it doesn't go on? "—"well I'm waiting a little longer." Kulik holds out the book for him. Word after word, he stutters toughly, tears in his eyes. The crammer chases the monkey:

"You have learned? Maybe, *you* believe that. But what have you done? You have drawn immortal verses into the dust. What should become of you?"

At this moment, the whole class, from the foremost to the back row, is convinced that B. has drawn immortal verses into the dust, and does not know what should become of him.

Another school bell is ringing.

In the next math lecture hour, Jäck just sketched another quadratic function nicely onto the board. He may sit down.

"Sylla come to the board and calculate the maximum." Sylla comes, differentiates, sets the derivative $y$-prime to zero, and calculates $x$. A mature achievement, one could say even worthy to Caesar. But Lina is still not saturated in her boring urge to know. She continues to interrogate, like complaining.

"Yes Sylla why is that the maximum now?" Sylla is an honest and open person and says:

> "I'll see that right away." "Sylla you are a bullhead."

At this moment, the whole class, from the foremost to the back row, is convinced that Sylla is a bullhead. Well, maybe in the end someone else discovered the law, I think. But no such thing does ever happen. And Sylla is a bullhead? The conviction overpowers me, too.

Another school bell is ringing.

During the break, Kulik asks:

> "Did you get what Lina explained in the last hour?" Torleß answers:

> "She thinks that if you can't calculate the third roots in the formula, they will become complex. But at the end of the whole calculation, the formula correctly gives the three roots of the third degree equation."

> Kulik: "The Lina says, Bombelli or such an Italian has discovered it."

> "Yes, yes; I know everything you say. But isn't there something very special about the matter? How should I express that? Just think of it like this: In the beginning, such a calculation contains very solid numbers that can represent meters or weights or anything else tangible, and are at least real numbers. At the end of

the calculation, a similar situation holds again. But the beginning and the end are connected by something that doesn't exist. Isn't that like a bridge, of which there are only the first and last pillars and which you still cross as safely as if it were completely there? For me, such a calculation has something dizzy; as if it were a part of the way God knows where to go. The really uncanny thing for me is the strength that lays inside such a calculation, and holds you so tight that you end up properly again."

Beineberg grinned: "You almost speak like our priest: '... you see an apple, —these are the vibrations of light and the eye and so on, —and you stretch out your hand, to steal it —these are the muscles and nerves that set them in motion. —But there is something between the two and brings one out of the other." "And that is the immortal soul that has sinned ... " He imitated the catechist in the way he used to produce the parable. "Yes, yes, none of your actions can be explained without the soul that plays on you like on the keys of a piano."

Then B. closed dryly: "By the way, I am not very interested in this whole story."

The knowledgeable listener knows how bad was Törleß fate. But in the end, he won't lose all sympathy. After having overcome the events of his youth, he later became a young man of a very fine and sensitive spirit. The past humiliation was that small amount of poison that is necessary to take the overly safe and calmed health of the soul and to give it a finer, sharper, understanding vigilance.

Another school bell is ringing.

But school itself never ends. Pfeiffer is still at school, too. He has to catch up with the high school, they have agreed, even if only for fun. On his first day, when asked, he gets up and says his name with a friendly smile.

"With one or two f?" "With three."

"Why with three?"—

"With however many, Pfeiffer, you will have to get used to discipline."

At this moment nobody in the class, from the first to the last bench, knows with how many f's the name P?eif??fer is written. The discipline is bad. Teachers are missing from the empire.

Minister Rust banned the film because "he ridicules the school."

Heinz Rühmann is alarmed and travels to the east. His aviator friend Oberst Angermund negociated for him. After a long-hour train journey, he is now within a triple ring of barbed wire. He is waiting. Those responsible prefer to watch the film by themselves. He reads a bit in the library. Afterwards, while walking, he discovers behind another barbed wire fence a small bent man, and next to him a German shepherd dog who follows and adapts to his steps. A few days later, Rühmann can hardly believe what he had experienced. From that day on everyone in the cinema can learn that Pfeiffer is written with three f's, one in front of the egg, and two behind the egg. And Heinz Rühmann smiles on the canvas.

The last school bell is ringing.

Dear friends: just consider this an end. All books that have inspired these lines are—or should have been banned once upon a time. How can my humble lines change anything about it? But they shouldn't prevent you from having the beer be tasty for you tonight.

## Ralph Vaughan Williams

In the audio book can listen to the Gavotte from *Suite de Ballet (1924)*

A great writers is quoted for consolation and as a warning, both for the reader and also for me.

He walked the path he had to take, a little careless and uneven, whistling to himself, looking side-ways with his head bowed sideways, and if he went astray, it happened, because for someones there is no really correct way.

—*Thomas Mann, Tonio Kröger*

## 2.4    The Not so Good Old Days

The memory embellishes all our past experiences, as already the Old Greeks did know.

"And if you don't want to listen to the songs from the old westerns, you can go to bed quietly and tomorrow morning drive the pigs out to the fields." You can already read this or something similar in Homer's Illyias.

During my school years, my classmate Kulik had the habit to entertain me with difficult problems from mathematics. In the earlier years, it was a geometric derivation of the binomial formula or the Pythagorean theorem, so later the determination of the focal length of a perspective drawing or a complicated special case to calculate an integral. Too, he was fond of fundamental questions, such as whether the coordinates were the parts of a point. As long as the teacher's directive was not yet in the room, no other classmate dared to think so far ahead.

On the ten-minute walk to the sports field, he keeps asking about my father's career. Now I just got another miserable grade on my German essay. For distraction, I talk about my mother. It was almost her advice for my grief. During her school career, she had got the habit to write *two* class essays during one lesson: during the same hour about two different topics with two different hand writings into two different booklets. First for a friend in her notebook, then the second essay in her own booklet. Later she became an actress. Kulik, too, finds all of this quite interesting.

Next time, on the way to the sports field, he talks about Stalin's hard upbringing with the Jesuits. The sports lessons are a repeated annoyance to me anyway, and my desire to do something more important arises. If not like Stalin, then what else? A week later he catches me in front of the entrance to the locker room. He needs help with the upcoming math exam. Just a piece of paper with the solution handed into the bank behind me, where Kulik is sitting next to Reinhold, —it would help. Kulik is completely underestimated by the English teacher, which language he speaks fluently, and nevertheless wrongly receives bad grades. Reinhold has repeatedly told me how Kulik gets stomach cramps before the math exams. For me, it looks like the perfect opportunity to make the world a little more just. Everything happens as agreed upon. Already during the class work, Reinhold whispers to Kulik.

"The result on Franz's note is wrong."

But Kulik believes my note more than his neighbor's hint. When the class work is returned, Lina shouts at me:

"Rother, I ask for an answer." (She always called me "Rother".)

I have to confess that I wanted to help Kulik.

Wittich approaches me after the hour.

"You should be aware of this: in the *Dritte Reich* you would be rid of your high school diploma."

A few days later, Kulik's school desk is empty. However, nothing happens on the part of the teachers. There is the rumor that he is with his uncle in London. He doesn't come back to school. A few months later, I did visit him with his uncle in London.

Wittich was my only classmate (but possibly not a female comrade) able to accurate judgments already in early childhood. Here is an example. The history teacher had just once again dutifully praised democracy in the Federal Republic of Germany as being perfect, like in heaven.

"In our state, the people determine the government through free elections."

Wittich objects. "I don't see anything like democracy."

"Why?" the teacher wants to know.

"I believe the government in Germany is acting according to the directives from the Americans."

Teacher Leßle agrees fairly loudly. "It is like that anyway, no doubt it is so in the end…" and the teacher tries to leave the impression that democracy still prevails.

Wittich also told me after the high school exams.

"That you are always the best one will be over quickly during your math studies, as you hopefully do realize."

Only he didn't say that the close community would be over, too.

# Chapter 3:
# Kafka and Cats

## 3.1   Reading Kafka: *A Country Doctor*

Hello Hans-Peter,

Please try if it works for you. I send you *A Country Doctor* from the CD with Gerd Westphal's Kafka recitation. It's really great. I've listened to it three times lately.

All the best Franz

Dear Franz:

Now concerning Kafka: of course, that's still hype, and I hope you know Kafka's letter to the father! (If not read or hear it immediately!!!) "The country doctor" is a wonderful surrealistic story, in which the black romance of Johann Heinrich Füssli (1741–1825, in England known as Henry Fuseli), even the Moses of Michelangelo ("the fingers in the beard") appear and which, on the other hand, feeds the later vampire films (Polanski's dance of the vampires, etc.). In addition, I realize during my latest reading a fundamental criticism of religion ("the pastor is sitting at home"), but also the criticism of belief in the modern (natural) science ("but the doctor should repair everything")...and—last but not least—not to forget: Kafka's fear of femininity or his own sexual drive (see the servant against Rosa—). But what interests me most are YOUR own associations. I think that would be great!

—Lghp

In the audio book version you may listen here to *The Moorish Cloth* from *Siete Canciones populares Espanolas*, by Manuel de Falla.

Why does fascinate me in this story? Especially the horses. They are a healthy sibling pair of spriting browns. Crapping out of a pigsty, they fly the earthly wagon all over the place. Curious and cheeky like a cat, both of them put their heads through the window and even tear it open. They are "sewn together" from different animal species.

Moreover, the breathless, furious chase during the whole night and fog campaign, which brings the poor doctor to the most extreme tension, amidst the incredible surprises. And in spite of that, everything seems credible and suitable for the rushed life of a country doctor. In the beginning, it wasn't more than that.

While reading once more, a lot of observations were added. The whole story begins to look fishy. Hans-Peter gives a mythological explanation that I would not have come up with by myself. But I want to point out the discrepancies without following a so far-fetched interpretation. The common people's doubts about religion and how their belief is replaced by expectations in modern medicine. The country doctor finds this attitude non-sensical or at least exaggerated:

> I am not a world improver and leave him alone. I am employed by the district and do my duty to the edge, to where it gets almost too much. Poorly paid, I'm generous and helpful to the poor.

> That's how are the people around me. Always ask the impossible from the doctor. They have lost their old faith; the pastor sits at home and tugs apart the chasubles, one by one; but the doctor should repair

> everything with his tender surgical hands. Well, I do
> it as you want: I didn't offer myself; if you use me for
> sacred purposes, I will let it happen to me; what else
> do I, an old country doctor, want—robbed of my maid?

Ultimately, the music also goes through this whole metamorphosis, with a thank-you chorus onto modern medicine. Believe it or not, something similar exists in an opera by Puccini.

Why does he keep thinking about his maid? Couldn't he since a long time have taken her as his medical assistant? His relationship with her seems good and familiar enough for such a task.

Already when he initially considers the patient to be healthy, it becomes evident that not everything is in good shape with the doctor. Soon afterwards, the patient also notices during the examination:

> My trust in you is very little. You're just shaken off
> from somewhere, don't arrive at your own feet. Instead
> of helping, you narrow my deathbed. I would prefer to
> scratch your eyes off.

And then the doctor is quite satisfied after having explained to the patient that he must die. One has to admit, in modern medicine, despite all successes, many things go wrong, too. But even more, is fishy. I guess, he is more a euthanizer than a doctor. The doctor sees and tells the patient that he wanted to hurt himself. In the end, the doctor and the sick both agree with this view and even find it good.

> I, who have been at so many a bed of the sick, far and
> wide, I tell you: your wound is not so bad. Created at
> an acute angle with two blows from the hoe. Many ones
> offer their side and hardly hear the hoe in the forest, let
> alone that it comes closer to them.
>
> "Is it really as you tell, or are you deceiving me in a
> fever?"

"It really is like that, take the honorary word of a doctor with you."

And he took it and fell silent.

But nobody seems to notice the slightest fraud. Why is the doctor getting out of the dust as quickly as possible? In the end, does he fear the discovery of his inability and, worse, of his wrong intentions?

In the audio book version you may listen here to *From Asturia* from *Siete Canciones populares Espanolas*, by Manuel de Falla.

## 3.2    The Neighbor

Dear Franz,

Kafka's story *The Neighbor* has been the subject of one of my weekend seminars with Sam Weber (1996). At that time we talked about the "unsecret" in Freud, an eerie state full of ambivalence. Indeed, the "secretly homely" or "homely secret" —by looking more closely always turns out to be an "unhomely secret", getting more and more scary. Compare Freud's corresponding article. The seminar stood under the impression of the death of Emmanuel Levinas, an important Talmud specialist and discussion partner of Jacques Derrida. A reference to Kafka appears in Sam Weber's 2010 essay, which I attach for you. Even if I suspect, uncannily said that you understand it even less than Sam Weber's *tertium datur*.

### 3.2.1   The Neighbor
*by Franz Kafka*, stories from the inheritance

My business lies completely on my shoulders.[1] Two misses with typewriters and business books in the lobby, my room with desk, cash box, consultation table, club chair, and a

---

[1]    The English title is *My Neighbor*.

telephone, that's my whole working apparatus. So easy to survey, so easy to handle. I am quite young and the business runs along for me. I don't complain, I do not complain.

Since New Year, a young man has rented on the spot the small, vacant apartment next door, which for too long, I have unluckily hesitated to rent. Too, a room with a lobby, but additionally a kitchen. —I could have used the room and the lobby—my two misses sometimes felt overloaded already, — but which purpose would the kitchen have served for me? This petty concern was to blame that the apartment has been taken away from me.

Now this young man is sitting there. His name is Harras. I don't know what he's actually doing there. On the door is written: "Harras, Bureau". I have made inquiries, I was told that it is a business similar to the mine. One cannot warn against granting a loan, because it is a young, aspiring man whose cause may have a future, but one cannot really recommend granting a loan, because at the moment there is apparently no fortune. The usual information that one gives when not knowing anything. Sometimes I meet Harras on the stairs, he always has to be in a hurry, literally he rushes past me. I haven't seen him yet closely, he has already the office key prepared in his hand, and he is immediately opening the door. Like a rat's tail, he is slipping in, and I am left standing in front of the sign 'Harras, Bureau', which I have read much more often than it deserves.

The miserably thin walls that betray the honest man but cover the dishonest one. My phone is attached to the wall of the room that separates me from my neighbor. But I only emphasize this as a particularly ironic fact. Even if it would hang on the opposite wall, one would hear everything in the apartment next door. I have taken the habit to avoid telling the name of the customer on the phone. But of course, it doesn't take much cunning to guess the names from characteristic, but inevitable expressions of the conversation. —Sometimes I dance, with the phone at my ear, driven by restlessness, on the tips of my feet around the apparatus, and yet I cannot prevent secrets from being revealed.

Of course, this makes my business decisions shaky, my voice trembling. What is Harras doing while I am on the phone? If I would like to exaggerate very much—but one often has to do it to gain clarity—I could say: Harras doesn't need a phone, he uses mine. He has moved his sofa to the wall and listens. Whereas, I have to get to the phone, when it is ringing. I have to accept the customer's wishes, make important decisions, carry out large-scale persuasions, —but especially involuntarily, I have to report to Harras through the wall.

Perhaps, he does not even wait until the end of the conversation. After the point of the conversation sufficiently informing him about the case, he rises himself and scurries through the city, according to his habit. And before I hung up the ear-cup, he is perhaps already working against me.

## 3.2.2 Comments from a Friend, and my Answers

> The important issue about the story *The Neighbor* is its relation to the DEATH which is always close to us as a scary guest (at the table, in the bed and at other occasions) and never allows us to be rejected. Such a symbol also appears in other places in Kafka's work, for example in the shape of *Odradek* in *The Care of the Housefather*. This "object" is also depicted in one of Jeff Wall's most famous art photos, my favorite photo artist.
>
> —Lghp

In the audio book version you may listen here to *Song* from *Siete Canciones populares Espanolas*, by Manuel de Falla.

> Hello Hans-Peter,
>
> Isn't this little story a wonderful parable for competition in business, as it is always lurking in the background; similarly in science, and in all types of working environments? If one looks at the story in this way, it gives a little comfort and amusement.
>
> All the best Franz

Remark (Another opinion). But the competition is only the obvious problem of the text, the real problem is a deeper uncertainty and anxiety. Hesitancy, pettiness, distrust, anxiety, self-blame and obsessions shape his existence. — The text shows the genesis of prejudice and paranoia.

> Dear Franz,
>
> Your approach to interpretation is justifiable, provided one applies it to the world of employment and the capitalist administration as a whole (see the work of sociologist Max Weber and above all the study by Siegfried Kracauer: The employees, 1929). Kafka was an employee of the workers' accident insurance company and a German Jew in Prague. With his illness, the overpowering loyalty to his father, and his crashed love affairs (especially Milena and Felicia Bauer) he was, so to speak, the successor to Herman Melville's *Bartleby*. Too, see Kafka's *Land Surveyor*, *Famine Artist*, *Josefine*. You always find him seeking for hope in writing. It was a life "directed towards death", as maybe Heidegger would have said. Neither being a valued citizen nor a really poor labourer, he was trapped between the classes of citizens and workers, as a foreign, alienated, part of the absurd life. Do you call that "amusing"? But for Kafka it was probably deadly serious: "Life … is only as long as the time you lose" I would say reflecting on his despair.
>
> —Lghp

In the audio book version you may listen here to *Polo* from *Siete Canciones populares Espanolas*, by Manuel de Falla.

> Hello Hans-Peter,
>
> When I say "amusing" for my response to this story, the cat hatched from the sack. It becomes obvious how comfortably I have already settled into my retirement perspective. Many years ago, I, too, was typewriting in a tiny office, or sucked too little useful knowledge out of my fingers. I probably wouldn't have told my

offspring that I am amusing myself with the paranoid competitive pressure that is shown as a caricature in this Kafka story.

—Greetings Franz

Dear Franz,

Thanks you again for your comments and your "joke!" And since I have just tried to clarify Sam Weber's article about *tertium datur* in my own words, I send it to you for review. Maybe you will discover something new in it? —Too, I find the problem with the cat exciting, because I am concerned with the problem of distinguishing animal and human; or with Jacques Derrida's words: "And if the animal could answer?" (Derrida was a passionate cat lover, too.)

—Lghp

## 3.3 My Cat

With the computer I copy the synopsis of the book,
which pretends to be so utterly important,
Meanwhile you tip over the glass of water
on the table with your front paw.
I cannot but laugh.
You retreat to the cardboard box in the corner,
turning your back on me.
A black stripe of black fur,
right and left I see, somewhat irregular, dazzling white.

When I look at the black spot on your right hind paw,
only the right one,
my love is renewed by itself.
You only eat the cheap canned food
and lick the carpet fibers.
You've been vomiting twice already.
The red spots on my arm,
will you bite again tomorrow morning?

## 3.4    The Cat Is Injured

My cat is still doing badly. She is just coming and settles herself on the keyboard, asking me what I'm doing here. She also wants to jump on my lap. I can feel that she has a fever again. I can now see the bad wound on the left hind leg. It still doesn't heal well. Everything happened last Tuesday. May be, she was so badly frightened by the road works that she jumped through a broken glass window into an uninhabited house in the neighborhood. Anyway, it has happened last Tuesday morning. In the evening I saw her again, but I still didn't know about her wound. That gave me a greeting by a bad scratch. For a week now, the animal has only been living in my house. If she were healthy she would have plagued me long ago, and meowed until I open the window, and she was gone again just for one day, sometimes two. When will such a life come back? Now her accident is eight days ago. After I come back from the merchant this morning, she has disappeared, but there was only a small gap open at the top of the kitchen window.

Yes, the cat came back afterward. She hadn't hurt herself again when she was squeezing herself through the narrow window gap. Usually I left a lower window open for her, and only stuffed in a pillow. Then she always came back at different times of the day or night. Finally, a badger or other clever predator noticed the trick, broke in at half past two at night and tore open several coffee bags. Otherwise nothing happened, the badger has simply disappeared.

In the audio book version you may listen here to *Sylvie*, by Gabriel Fauré.

# Chapter 4:
# Mozart and Bach

## 4.1   Zerlina and Masetto in Italy
*In the Footsteps of Mozart's Don Giovanni*

*not by Lorenzo da Ponte*

THIS SHORT SCENE IS the result of viewing many performances of several Mozart operas. During the radiation therapy because of cancer, I was rather tired and had enough idle time. Too, the ideas reveal how feverish I was at that time.

Zerlina: My love, get up. The journey is long!

Masetto: What's there to eat where we're going?

Zerlina: We'll see. We've been recommended a first-class restaurant in Bologna. There are supposed to have many specialties and even something very special for sightseeing.

Masetto: Do they serve potato salad with sausages?

Zerlina: Good heavens, you can already see the building in the distance. It looks like it's made from sugar.

Masetto: I hope it's not so dark inside. I'd like to sit out-doors.

Zerlina: You're a guest here and behave accordingly.

Masetto: Oh, if only I were at home in my dog's kennel, or if I could feed my cats.

Zerlina: Look over there in the corner! They say that anyone who touches that beautiful marble statue will be turned into stone, unless he has been without any sins all his life long.

Masetto: The white spoils my eyes. I need a spray can to give it some color.

Zerlina: Let's behave like guests, because that's what we are here.

Masetto: Waiter, potato salad and sausages, please!

Waiter: Excuse me. Today we're only serving the dish of the day.

Masetto: What is that? Why are the plates triangular?

Waiter: My dear friend, in this world, not everything that is square is a circle.

Masetto: I immediately will throw the strange plates at the boy. And why does he always sing at the door: no, no, I won't serve you anymore.

Zerlina: Behave yourself! Customs are different all over the world.

Zerlina: Waiter, please, what is today's specialty?

Waiter: Beef goulash with potatoes and green beans. I can only recommend it.

Masetto: Waiter, I only have a fork. And do I have to cut the goulash with my pocket knife?

Waiter: If the guest has a question, we understand each other better. I'll bring you a spoon right away.

Zerlina: Look how helpful he is. He can safely touch the statue in the corner.

Masetto: As soon as I've given it a new coat of paint, I'm sure he won't be able to touch it again. By the way, what's the waiter's name?

Zerlina: They call him Leporello. But the restaurant's owner is even more famous than him.

Masetto: That must be a strange guy. Why are there so many small rooms up there?

Zerlina: That's none of your business. Don Juan is the master of the house, you're just his guest.

Masetto: How can Don Giovanni be the master of the house when he's long dead and gone to hell?

Zerlina: Then his heirs are the masters of the house. As famous as he was, and how many recordings there are of his famous opera, there is certainly a rich community of heirs.

Masetto: Don Giovanni has raped countless women. He even has murdered, I know that. How can such a villain be so famous? I'll smash the statue and throw the chunks at the landlord's head!

Zerlina: You'll be turned to stone yourself first.

Masetto: At least let me try. And then we'll storm all the rooms up there together.

Zerlina: Stop it now and behave yourself or I'll never go on vacation with you again. While I'm spending my money I want to have a good time. Waiter, please bring the dessert.

Waiter: I'm sorry, dessert will only be served late in the evening, after the tour of the entire castle has been completed.

Zerlina: Is the visit free, or how much does it cost?

Waiter: That depends on how long you want to stay upstairs.

---

Masetto: I would definitely be bored without Mozart's music.

Waiter: You can even play it.

Zerlina: But really Mozart's music?

Waiter: No, ladies and gentlemen. Today, most people no longer like this music. They say it no longer inspires them and even disturbs them.

Masetto: How stupid people are.

Leporello: Listen, I can give you a little comfort. Once a month we organize something very special: a big garden party. That's when lots of famous singers perform. They continue to be fond of Don Juan's serenade, and the ladies from upstairs show themselves at the windows.

Masetto: But I ask you, are there enough rooms for all his mistresses? Besides, Leporello, you have kept a list of his mistresses.

Leporello: Yes, I still have the list, it really is very long. Many girls and women, from all countries, from the countryside and from the city, peasants and aristocrats, they were all Don Giuovanni's mistresses.

Zerlina: And do they all live up there now?

Leporello: No, of course not. It is a special honor for a selected few ones to be living up there. And to be allowed to come to the party is even more difficult, and only very few get permission.

Waiter: I can tell you in confidence that some of the ladies take the opportunity and are very revealing.

Masetto: Watch out! Do you still remember, Zerlina? After one of these incidents, I hurried to see what was going on. And I had even my gun with me. Leporello came and beat me up. I screamed for help and you, my love, came to rescue me. It was only because you had a very special glue with you that you were able to save me quickly.

Zerlina: Oh, what are you fantasizing about glue. By the way, you know very well that it wasn't our dear waiter.

Masetto: That's right, Don Juan had disguised himself and had beaten me. But nobody did realize it until only much later.

Zerlina: Leporello, tell me, do you still have the big black hat, the wide cloak and the sword you wore as a disguise?

Leporello: These items all still exist. But even I'm only allowed to wear them once a month on special occasions.

Masetto: And when will we get to see all this?

Leporello: Unfortunately, the next demonstration isn't for another week.

Zerlina: We'll stay until then.

Masetto: But in the meantime, the statue can dine with us and sing for us.

Zerlina: What would the statue like to have for dinner?

Masetto: I'll give it my cat food. It tastes heavenly to my cats!

Leporello: We keep asking the commander. Does he want to dine with us? And then he nods his head, just like I'm doing now. But then he tells that he can only eat heavenly food. He can't digest earthly food.

Masetto: My cats think their food is heavenly.

Leporello: Don Elvira has traveled far to find this heavenly food. She sent all the angels up and down the stairs of the high towers of San Giaminiano, like the flutes in Mozart's music make it sound. But she only found a few bats.

Leporello: Donna Elvira, in particular, has become deeply saddened and still is hurt by how everything went wrong back then. Despite all his evil behavior, she continues to feel sorry for Don Giovanni. Right up to the last minute, she kept wanting him to become a better person. But that couldn't work out.

Zerlina: Yes, I remember the wonderful aria in which she expresses her contradictory feelings. It sounds like a Bach cantata, but it is also Mozart's music.

Masetto: Mozart's music is too easy for children and too difficult for adults.

Leporello: Back then, I simply put an end to my relation with the Don, and immediately took up service with a better master. Time can heal a lot of suffering, but only if one is ready for it.

Zerlina: On the other hand, there must be signs and warnings.

Leporello: Many people hear the sound of the trumpet from the belly of the statue, and are always terribly frightened.

Zerlina: Mozart also uses the trombone, after Don Giovanni had made too much fun of the old man.

Masetto: I'll going to check if I can find the trombone inside the statue.

Zerlina: But don't touch the statue.

Masetto: I just want to have a look.

Leporello: Yes, that happens all the time, especially when the guests are having a good time late in the evening. The trombone choir then resounds loudly through the hall.

Masetto: Today the statue is supposed to sing something else for us. I would like to hear it sing: All the birds are already here. That's part of spring and good weather.

Leporello: Ladies and gentlemen, bear in mind that the old man may not be able to sing such funny songs.

Masetto: That doesn't matter. We'll teach him. I have written it down here. Please read it.

Leporello: Unfortunately, I can't read by moonlight, I have not been taught that.

Zerlina: There's enough light here. Read on.

Masetto: All the birds are already here,

> I want to catch many,
> because I am the bird catcher,
> well known to old and young.
> Blackbird, thrush, finch and titmouse,
> are singing loudly all,
> Yes, I can see the starling too,
> Good heavens, there's even a couple.
> I am the bird catcher,
> I have many birds in my cage,
> and many girls, too.

Leporello: The case amazes me. The old man has to learn a new role from a completely different opera.

Zerlina: This other opera is by Mozart, too. He'll manage it all right.

Masetto: That makes the old statue really young again. He's already bored of just playing the trombone.

Leporello: I do everything for my guests!

Zerlina: Thank you very much, dear waiter. And you, my darling, you are magnificent. I'm already looking forward to my next vacation with you.

## Epilog

Masetto: "...What is there to eat where we are going?

...Is there potato salad with sausages?"...
And I recommend an "appelwoi" to go with it!

Greetings from Hans-Peter

Dear Hans-Peter:

That is possibly the right recommendation. Masetto is a really rural type. I can imagine that his taste for dinner is like that.

The good fortune that he is taking the village beauty Zerlina as his wife is thoroughly spoiled by Don Giovanni. At every opportunity, the Don makes the most sweet proposals to Zerlina. Although she is not one hundred percent averse, she keeps him sufficiently in suspense. Only the somewhat simple-minded Masetto doesn't see that.

During the second act, after the Don and his servant have swapped their outer garments, the Don finds his opportunity for revenge. Disguised as Leporello, the Don organizes a paramilitary action, to capture Don Giovanni. Two groups are to go out in different directions after Don Giovanni. But he (the Don disguised as Leporello) remains on the spot alone with Masetto, to (supposedly) give the Don the rest. While Masetto is still boasting about his two sharp weapons, the Don grabs one of them and beats Masetto up. He calls Zerlina for help. She arrives and heals him in a very sweet manner.

The Zürich performance shows all this theater most vividly. The Swiss penchant for military games was helpful to put them in scene. In the opera's epilogue, all the victims make it clear how their lives are to continue after Don Giovanni's has been going to hell. Finally, Zerline and Masetto are the happiest ones. They simply want to go home and dine comfortably together. The eternal coffee and chocolate and ham from Don Giovanni's grace is certainly no longer their taste.

How about potato salad and sausages and cider?

Zerlina a nd Masetto in Italy
In the Footsteps of Mozart's Don Giovanni

**Franz Rothe**

## 4.2    Johann Sebastian Bach's Futile Journey to Fame

There are two excellent movies about the old days of Johann Sebastian Bach.

The movie *Johann Sebastian Bach, part IV The order of the stars* by Ulrich Thein,1983-84 gives a rather comprehensive account of the old Bach's life: the composer, the father of a big family, and the old man suffering from eye cataracts.

The movie by Victor Vicas from 1980 shows us the efforts of Johann Sebastian Bachs to get a position at the court of Friedrich II, his journey of 1747 to Berlin and Potsdam, and his return to Leipzig.

Bach's Christmas Oratorio was his greatest success. It was made possible only thanks to the support of the newly appointed rector of St. Thomas's School who was well disposed of him.

Three years later, this rector Gesner was appointed as a professor at the University of Göttingen, and with the new rector Ernesti the old difficult and disappointing times returned for Bach. The newly appointed rector soon prevented a second performance of his Johannes Oratorio on flimsy grounds. In addition, his eldest son Friedemann was not satisfied and successful as an organist in Halle.

These were probably the reasons for Bach's decision to travel to Frederick the Great's court in Berlin. His second eldest son Karl Phillip Emmanuel had become the leading conductor at the court. During these years, Frederick the Great decided to turn Berlin into a globally respected cultural center. But Frederick had several faces: for him, culture was only ever intended to be a glamorous accessory. Ultimately, the military was vital. The Seven Year War was still coming soon.

Bach wanted to listen to the court orchestra in Berlin, wanted to play several organs and dedicate a significant composition to the king. Through all these endeavors, he hoped to gain court positions for himself and his eldest son Willhelm Friedemann. The movie impressively describes the harsh realities of life around 1750, especially during Bach's and Friedemann's arduous journey from Leipzig to Berlin. The entire traveling group has to clear fallen trees from the path along the way. The princely cavalry's hunt for poachers interrupts the journey. Poor Jews want to travel to Berlin but cannot afford transportation. Deserters are caught and cruelly punished.

Following the epic description of Bach's journey, we see his vain efforts to gain the king's approval. When he finally succeeds, he is immediately questioned about whom he intends to bring with him from his family if he is hired. Finally, he is allowed to see and hear the king and his orchestra, and is then requested to play on the cembalo. He is soon interrupted while playing the harpsichord in French style. He is now asked to improvise a fugue on a theme Frederick the Great has intonated for him on the flute. After Bach improvised a three-part fugue, the king demanded for a six-part fugue. Bach asks for time to think about it until the next day. In the meantime, a famous ballet dancer has arrived for rehearsals. The king only has eyes for her art, forgets Bach and barely pays attention to his six-part fugue. The journalists are also all too critical of Bach. Bach departs without confirmation for himself or his son Friedemann.

During the journey back, he begins composing the "Musical Offering". This includes the valuable trio sonata for flute and violin, and thirteen fugues and variations on the royal theme. Returning to Leipzig, he sent a copy of the composition with a detailed dedication to the king in Berlin. He does not receive any compensation for his major original compositions.

His most significant further compositions from these last years are the Goldberg Variations, which are still nowadays black bread for all famous pianists; the Mass in b-minor and the Art of Fugue. After two unsuccessful eye operations, Bach died from medical complications. His five other sons and daughters almost all became impoverished. Elisabeth marries the loyal employee Altnikol, and still has to accommodate and care for the mentally handicapped son Gottfried Bach for the rest of his life. The movie shows a harshly realistic picture of the last years of the greatest composer. It also shows the general life circumstances under which Bach had to live and work.

The movie gave me some references to my own life, too. I have played the flute sonata and other pieces from the Musical Offering on the flute myself. I still remember well how I switched the record player on, and played along with the flute part. My old father listened with his small dog on his lap. That was during a visit home for Christmas in the last years of my father's life after my mother's death.

Bach's eye condition was cataracts: an age-related clouding of the eye lens. This was and still is nowadays a common disease, especially among older males. The famous mathematician Leonard Euler also went blind by this disease. However, he did not undergo surgery, instead he decided to only research elementary mathematics, and was assisted by his son. Euler did not get on well with Frederick the Great neither, who had called him a "mathematical cyclops". Euler moved to the court in St. Petersburg. If only Bach had followed a similar path!

Nowadays, cataracts can be treated with surgery. This was a surgery I underwent myself. The encrusted eye lens is pulverized and suctioned out in seconds using ultrasound. Afterward, an artificial lens can be inserted. The patient recovers within a few days and can see clearly and work better again.

I still remember quite clearly the mischief that happened to me the next day after surgery, by my fault. Like I was used to, I drank some red wine the very next day, and unexpectedly fell down the stairs at home. All I got from falling down the staircase was a bruise on my forehead. With some makeup, I was able to conceal the injury so well that neither the students nor the doctor during the follow-up examination noticed the bump. Since then, I have gladly avoided alcohol.

The movie about the old Bach brings back more memories for me than the other much more optimistic Bach movies. You really have to be grateful for advances in medicine. Many older people have been able to live a better life thanks to them.

# Chapter 5:
# Musical Activities

━━━━━━━━━━━━━━━━━━━━━━━━ ● ● ●

SINCE CHILDHOOD, I HAVE become more and more familiar playing the flute. The following early recordings were made as my father just had bought a tape recorder. The poem *Apfelkantate* by Hermann Claudius we had learnt at school. The poem describes the growth of an apple over the seasons. I liked it so much that I composed a song for it.

> Der Apfel ist nicht gleich am Baum,
> Da war erst lauter Blüte,
> Da war erst lauter Blütenschaum
> Da war erst lauter Frühlingstraum
> Und lauter Liebe und Güte
>
> Dann waren Blätter grün an grün,
> Und grün an grün nur Blätter.
> Die Amsel nach des Tages Mühn,
> Sie sang ihr Abenlied gar kühn
> Und auch bei Regenwetter.
>
> Der Herbst der macht die Blätter steif,
> Der Sommer muss sich packen.
> Hei, daß ich auf dem Finger pfeif,
> Da sind die ersten Äpfel reif,
> Und haben rote Backen.

———

Und was bei Sonn und Himmel war,
Erquickt nun Mund und Magen,
Und macht die Augen hell und klar.
So rundet sich das Apfeljahr,
Und mehr ist nicht zu sagen.

Here is a somehow awkward translation:

The apple is not immediately on the tree,
First there were all blossoms,
There was all blossom's foam
There was all spring's dream
And all the love and kindness

Then leaves were green on green,
And green on green only leaves.
The blackbird after the day's toil,
He sang his evening song boldly
And even in rainy weather.

Autumn turns the leaves stiff,
Summer must clutch itself.
Hey, I'm whistling on my finger.
The first apples are ripe,
And have red cheeks.

And what was in sun and sky,
Now refreshes mouth and stomach,
And makes the eyes bright and clear.
Here's how the apple year ends,
And that's all there is to say.

In the audio book follows a recording, sung with my high voice.

As I began to realize that the old composers were using the musical notes and phrases like building blocks, and not every phrase needs to sound as if it were part of a song, I felt that I should try myself. In the audio book you may listen to my origin first result.

For a small concert that I had organized in the church of Wolfartsweier, a village near my home town Karlsruhe, I have performed together with Marlies Schubert Händel's German aria *Meine Seele hört im Sehen* ("My soul hears in seeing"), HWV 207. Unfortunately, this is the only recording of the many musical activities during these years of my youth.

## 5.1     The CDs with Pianist Thomas Turner

I have continued my musical training in North Carolina. During the years 1992–2003 a repertoire was practiced together with the pianist Thomas Turner, and three CDs have been recorded.

See the covers for *Withered Flowers*, *Serenade* and *Mandoline* shown on the pages below. These recordings were possible through the vigorous encouragement of Thomas and pianist Heather Coltman as well as flutists Irene Maddox and William Bennett.

### Claude Debussy

In the audio book you have heard *Romamce* from the "Eight Early Songs", played with alto flute and piano.

These were golden years for the musicians Rothe, and maybe Thomas, too.

When I now put together an album with two more CDs from my own musical activity after so many years, it is a very nostalgic endeavor. I feel myself imitating the old boarding director Sesemi Weichbrodt from the *Buddenbrooks*, who nicely repack-aged her old stuff for Christmas and gave it away again. For me with my poor eyesight, all flute playing is an activity from the past, as disintegrated as the *Buddenbrooks* family of Thomas Mann.

Nevertheless, I would like to invite you to listen to my last package of two CDs, which was released in November 2019.

*A Bouquet Inspired From Songs and Instrumental Music*

Today from a distance, this collection from the golden years of Rothe and Turner amazes me with its forceful and accurate performance. Professional suppleness is almost achieved. The collection includes some almost forgotten pieces and surprises with (of course, less well-known) own transcriptions.

### Beethoven, *Rondo in G major, WoO 41*

In the audio book you can hear my own transcription from violin to flute.

### Beethoven, *Serenade for flute, violin and viola, Op. 25*

This chamber composition was definitely finished by late 1801 when Beethoven offered it the publisher G. Cappi. In 1803, Franz Xaver Kleinheinz arranged the serenade for flute (or violin) and piano. Beethoven checked and approved this arrangement and it was printed as his Op. 41. In the audio book I have included the very lively short movement. *Allegro scherzando e vivace.*

### Schubert Songs

The songs *Standchen* (also called Serenade) and *Das Fischer*mädchen (The Fishing Maiden) are number 4 and number 10 of Schubert's cycle *Schwanengesang*, D 957. The arrangements for the flute were composed by Theobald Böhm, the designer of the modern flute. They are virtuoso playful fantasies about these songs. In the audio book is included the famous Serenade.

### Eugène Bozza, Aria for Alto Flute and Piano

This piece is already originally available in several versions: in a very high pitch for violin or flute and in a lower pitch for clarinet, or even for alto saxophone. Under the keyword "Bozza Aria" you will find on the internet many recordings on the most varied instruments. I transposed the piece myself into a pitch for alto flute. This version is included in the audio book. In addition, I have produced a second version for flute in a more convenient pitch that can be heard here, too. I have rewritten the piano part with the aim of making it easier to play.

## Robert Baksa

Robert Baksa was born in New York City in 1938 and is of Hungarian descent. He is one of the most productive American composers. One of his earliest works for flute is the Flute Sonata Number 1. It was created in 1976 but was only premiered in the late 1980s. It is among the winners of the 1994 "Newly Published Music Competition", —a prize donated by the "National Flute Association".

The movements are: Allegro, Cadenza I, Adagio, Cadenza II, Allegro

## Beethoven, *Rondo a Capriccio in G, Opus 129*

Beethoven composed this hit in 1795. The piece rightly has the nickname "Rage Over a Lost Penny". This fiery composition made me have a lot of fun playing. Thomas still writes to me: *familiar*

> Franz, the two CDs arrived yesterday. Thank you very much! I haven't listened to them yet, but I'll do it soon. In any case, I think you should clarify *who* is playing Beethoven's *Rondo a Capriccio, Opus 129*. This is the only way the listener can know for sure. I assume you play it, but of course, it could be the second flautist, who knows? You really put some of our best pieces on the CDs!

## Francis Poulenc

Sonata for Flute and Piano: Allegro malinconico, Cantilena, Presto giocoso. Here is an early, not perfect, own recording Poulenc95

Of Francis Poulenc's *Flute Sonata* there exist many excellent recordings, starting with the original from 1957, played by the world famous flutist *Jean-Pierre Rampal* and the composer.

## Benjamin Godard, *Legend pastoral* from the cycle "Scotch Scenes"

Benjamin Godard Legende Pastorale

## Robert Schumann

The three movements of the fantasy pieces, Opus 73, are already by Schumann titled in German.

> *Zart und mit Ausdruck* (Delicate and with expression);
> *Lebhaft, leicht* (Lively, light);
> *Rasch und mit Feuer* (Quick and with fire).

Robert Schumann has written the fantasy pieces Opus 73 in a version for cello and piano, —as well as a version for clarinet and piano. For my own adaptation for flute, I only had to choose the appropriate octave in the melody part.

## Schumann 2

The sharper articulation of the syncope at the beginning of the third movement, which is more suitable for the flute, is taken over by Schumann's cello version.

Robert Schumann op. 73 Rasch und mit Feuer

## 5.2 My Adaption with Two Flutes

The concert piece *Odelette* in D major, Opus 162 (in English "Small Ode"), is a late work by Camille Saint-Saens (1835– 1921). It was composed in 1920, and was originally set for flute and orchestra. The piece has a clearcut structure and is based on a melody of classic, even Greek flavor. The flute part is rich in playful and virtuoso ornaments. I got to know this piece around 1998, during the years when I played regularly together with pianist Thomas Turner. The simple and at the same time playful character of the piece did attract me immediately.

While the flute part is demanding, putting everything together seemed to be rewarding enough and within our abilities.

But Thomas was very reserved, especially because of the transcription for piano. This transcription contains several passages with only a single melody line for the piano. This is very unusual and awkward for any pianist since it does allow the piano sound to develop naturally.

Instead of insisting on practicing the flute and piano version, I set about rewriting the piece for two flutes and piano. The simple classic style of the piece gave me confidence in such an endeavor. In the version for flute and piano, there already exist variations with the character of a duet. On top of that, the long virtuoso runs were much easier to master by distributing them on two flutes. These properties of the composition make the transcription for two flutes a natural undertaking. Of course, there did arise additional problems during this endeavor. Indeed it became necessary to invent some additional parts of figured bass and to redistribute voices. After a few months of such work, I managed to get a transcription. We have played it together with Andrea as the second flautist and Thomas as pianist.

## Camillie own

The manuscript has been accepted by a music publisher, and is now available from allflutesplus, and: justflutes. In addition, there exists now a recording with two first-class flutists on their CD.

*Luminance* by Lisa Friend, Anna Stokes, Mark Kinkaid Released 2014, Champs Hill Records.

The version for two flutes and piano from the CD *Luminance* can be heard online on YouTube, too.

In memory of my (small) contribution, I could imagine appearing in the background of the cover picture. For example, as a dwarf sitting on a cloud, while I watch from a distance how the clear flute tones, regularly like raindrops from two clouds, swell from the mouths of the two nymphs.

## FRANZ ROTHE, FLUTE and THOMAS TURNER, PIANO
*"Withered Flowers"* Music for flute and piano

### LUDWIG VAN BEETHOVEN
(1770-1827)

1. **Rondo in G major**, WoO 41 [4:04]
*(arranged for flute by F. Rothe)*
Allegro

### MAURICE RAVEL
(1875-1937)

2. From the Suite "Ma mère l'oye" [3:50]
*(arrangement by Th. Turner)*
Pavane de la belle au bois dormant
Petit poucet

3. **Pavane pour une infante défunte** [5:03]

### GEORGE-ADOLPHE HÜE
(1858-1948)

4. **Sérénade** [1:50]
Allegretto leggiero

### CAMILLE SAINT-SAËNS
(1835-1921)

5. **Romance** [4:31]
Moderato assai

### ROBERT SCHUMANN
(1810-1856)
**Three Romances**, op. 94

6. Nicht schnell Moderato [3:15]

7. Einfach, innig Semplice, affettuoso [4:02]
8. Nicht schnell Moderato [1:01]

### RALPH VAUGHAN WILLIAMS
(1872-1958)

9. **Suite de Ballet** (1924) [6:03]
Improvisation Andante-Poco piu mosso-Andante
Humoresque Presto
Gavotte Quasi Lento
Passepied Allegro vivacissimo

### FRANZ SCHUBERT
(1797-1827)

10. **Introduktion und Variationen**, op. 160 [17:41]
über "Ihr Blümlein alle" aus den Müllerliedern
Introduktion Andante
Thema "Trockne Blumen" Andantino
Variationen I-VII

TOTAL PLAYING TIME: [54:46]

STEREO [D] [D] [D]
Ⓟ Ⓒ 1998 Franz Rothe
Recording: Charles Vaughn
Photos: Luz Maria Aveleyra and Franz Rothe
Piano: Steinway & Sons
Flute: Johann Hammig, Freiburg i. Br.

# Serenade

## Music for Flute and Piano

Franz Rothe, Flute

Thomas Turner, Piano

---

*"Serenade"* Music for flute and piano
FRANZ ROTHE, FLUTE and THOMAS TURNER, PIANO

CHRISTOPH WILLIBALD GLUCK
(1714-1787)

1. From the Opera "Orfeo" [4:30]
   The Dance of the Blessed Spirits

LUDWIG VAN BEETHOVEN
(1770-1827)
**Serenade** op. 41

2. Entrata  Allegro [3:16]
3. Tempo ordinario d'un Menuetto [3:45]
4. Allegro molto [2:15]
5. Andante con Variazioni [5:39]
6. Allegro scherzando e vivace [1:57]
7. Adagio, Allegro vivace e disinvolto [5:42]

BENJAMIN GODARD
(1849-1895)

8. From "Scotch Scenes" [5:30]
   Legende Pastorale  Andante quasi adagio

GABRIEL-URBAIN FAURÉ
(1845-1924)

9. Les Berceaux  Andante [2:46]
10. Sylvie  Allegro moderato [2:56]

ROBERT SCHUMANN
(1810-1856)
**Fantasy Pieces,** op. 73
*(transcription for flute F. Rothe)*

11. Zart und mit Ausdruck [3:09]
12. Lebhaft, leicht [3:33]
13. Rasch und mit Feuer [4:28]

TOTAL PLAYING TIME: [49:30]
STEREO  D D D
℗© 2000 Franz Rothe
Recording: Charles Vaughn
Piano: Steinway & Sons
Flute: Johann Hammig, Germany

"*Mandolin*" Song Transcriptions
Franz Rothe, *Flute* Thomas Turner, *Piano*

## "*Mandolin*" Song Transcriptions
### Franz Rothe, flute and Thomas Turner, piano

**FRANZ SCHUBERT** (1797-1827)
*transcriptions for flute by Theobald Boehm*

1. Serenade — [4:50]
   *Ständchen* Moderato
2. The Fishing Maiden — [2:26]
   *Das Fischermädchen* Allegretto
3. The Carrier Pigeon — [3:50]
   *Die Taubenpost* Andante con sentimento
4. The Linden Tree — [3:47]
   *Der Lindenbaum* Moderato

**GABRIEL FAURÉ** (1845-1924)

5. The Cradles — [2:43]
   *Les Berceaux* Andante
6. Sylvie Allegro moderato — [2:54]

**CLAUDE DEBUSSY** (1862-1918)

7. Romance Andante — [1:55]
8. Beautiful Evening — [2:16]
   *Beau soir* Andante ma non troppo
9. Flowers of the Grainfield — [1:52]
   *Fleur des blés* Andantino moderato
10. Here is Spring — [2:16]
    *Voici que le printemps* Andantino
11. Mandolin Allegretto — [1:25]

**MANUEL DE FALLA** (1876-1946)
**Siete Canciones populares Españolas**
*transcription by Thomas Turner*

12. The Moorish Cloth [1:21]
    *El paño moruno* Allegretto vivace
13. Seguidilla from Murcia [1:10]
    *Seguidilla murciana* Allegro spiritoso
14. From Asturia [1:59]
    *Asturiana* Andante tranquillo
15. Jota Allegro vivo [3:03]
16. Lullaby [1:32]
    *Nana* Calmo e sostenuto
17. Song [1:05]
    *Canción* Allegretto
18. Polo Vivo [1:29]

TOTAL PLAYING TIME: [42:35]

℗© 2003    Franz Rothe
Recording:    Charles Vaughn
Cover photo: Alice Mayer
STEREO  [D] [D] [D]

# A Bouquet *inspired from songs and instrumental music*

Franz Rothe *flute and piano*, Thomas Turner *piano*, Andrea Hauk *flute*

Beethoven, Schubert, Fauré, Godard, Debussy,
Schumann, Saint-Saëns, Bozza, Poulenc, Baksa

Franz Rothe *flute and piano*,
Thomas Turner *piano*,
Andrea Hauk *flute*

## A Bouquet Inspired by Songs

### Franz Schubert (1797-1827)
1. Serenade *Moderato*
2. The fisher maiden *Allegretto*

### Ludwig van Beethoven (1770-1827)
3. "To the far beloved" op.98 (1816)

### Gabriel Fauré (1845-1924)
4. The cradles *Andante*

### Eugène Bozza (1905-1991)
5. Aria (1971) *alto flute*

### Claude Debussy (1862-1918)
6. Romance *Andante alto flute*
7. Here is spring *Andantino*
8. Mandolin *Allegretto*

### Robert Baksa (*1938)
Flute sonata (1976)
9. *Allegro Cadenza I Adagio*
10. *Cadenza II Allegro*

## A Bouquet from Classics

### Ludwig van Beethoven
1. Rondo in G-major WoO 41

### Benjamin Godard (1849-1895)
2. Legende Pastorale from "Scotch Scenes"
   *Andante quasi adagio*

### Ludwig van Beethoven
3. Rondo a capriccio in G, op.129
   "The rage over the lost penny"

### Camillie Saint-Saëns (1835-1921)
4. Odelette in D-major, op.162

### Robert Schumann (1810-1856)
Fantasy pieces, op. 73
5. tender and expressive
6. lively, light  7. vivid and with fire

### Eugène Bozza (1905-1991)
8. Aria (1971) *flute*

### Francis Poulenc (1899-1963)
9. Sonata for flute and piano (1957)
   *Allegro malinconico Cantilena Presto giocoso*

## 5.3     My Completion of Bach's Flute Sonata BWV 1032

From the preface of the Leipziger Bach edition, one can learn the history of the flute sonata in A major, BWV 1032. The autograph has as a title, written by J. S. Bach himself:

"Sonata a 1 Traversa e Cembalo obligato di J. S. Bach." Besides the sonata, this autograph contains a concerto for two pianos accompanied by a quartet. The whole is written in that peculiar manner found often in Bach's cantata scores. On 15 pages, the concerto takes the upper 16 systems, while the three systems left at the bottom of the pages contain the sonata written in parallel. Only after the end of the concert, starting from the 16th page, does the sonata fill entire pages.

The first eight pages have been completely preserved. But from the following six pages, the three systems at the bottom, which would contain the first movement of Bach's sonata, have been cut away. Only page 15 is complete, again, and contains the last two measures of this movement. Because the ratio of the missing part to the existing part (the first 62 measures) is 3 to 4, one can guess that 46 to 48 measures are lost by truncation. The current autograph owner is Mr. Grasnick from Berlin.

He acquired it from the inheritance of Mr. C. v. Winterfeld, who, in turn has bought it a long time ago at an antiquary in Breslau for a few dimes. According to a message of Mr. von Senfft, the autograph seems to have been incomplete already at that time.

The editor of the Leipziger Bach edition tells: we are sorry to say that all efforts to complete the missing part were in vain, because an older, complete copy could not be found. Our edition contains the first movement fragment as an appendix.

To complete the first movement fragment is a real challenge suitable for flutists or keyboard players. Several completions have been published. I looked carefully at the version by Alfred Duerr, published by Bärenreiter, and the version of William Bennett, published by Chester, and have learned about still further versions.

Until measure 77, everyone seems to agree on how to complete the piece. But starting at that measure, all the versions differ significantly from one another, to an astonishing degree. Here are some ideas that guided me in my own completion attempt. My motivation to complete the piece comes from its valuable music. From the beginning I did not attempt to get an exact historical reconstruction, but tried to make really joyful music. The alternation of trio and continuo accompaniment brings a concertant quality to this joyful, upbeat piece.

Next I looked at the overall architecture of the piece. To the main entrance motive, so well tailored to the keyboard, Bach adds a jubilant flute call as a contrast. The next section brings long ascending arabesque lines and interesting modulations. The middle section recalls the entrance motive in a more playful way, and develops it into a long three-voice fugue. Here the fragment ends.

How can the piece go on? I guess the three main ideas must be recalled in reverse order, in enhanced and refreshed quality. Given the length of the gap, I think it to be appropriate to bring in some own ideas, but with caution. I have tried to keep my own small contribution as close as possible to Bach's manner, following the most strict composition rules.

It took my full attention from Christmas day 2003 until Valentine's day 2004 to come up with a satisfactory version. For their support, I have to thank my elder son Tilman, and Dr. Philip Burgess, Organist/Choirmaster at St. Luke's Episcopal Church in Salisbury.

myBWV1032

# Chapter 6:
# My Old Love for Mathematics

BEFORE GOING INTO THE middle of this difficult subject, here is given a rather (too) ironic view on the difficulties of teaching mathematics. Luckily enough that mischief did not happen to myself. But my former students would say, I was rather too close.

## 6.1   Become a Plumber!

Once upon a time, there was a professor of mathematics. He was doing quite well, except for a little something that bothered him at home. All the water taps dripped and dripped. He had already wasted many a free hour trying to remedy the defect himself, now he was fed up and called a plumber. The handyman was able to remedy the defect very quickly, to the satisfaction of the professor. At least until the plumber finally scribbled some almost illegible characters and numbers on his notepad, and handed the note over to the professor. He looked at it and was horrified.

"That's a third of my monthly salary."

"Lord, then they come to us and become plumber." The professor found the proposal worth considering.

"I just have to tell you one more thing. The masters with us do not like educated apprentices, and often do not take them at all. So it's best to say that you only have seven years of elementary school."

With determination, the professor goes to the master. And lo and behold, the trick with the seven years of elementary school works. Now he goes to the plumber apprenticeship three times a week. He can already cut threads and even measured the thickness of a pipe correctly. The journeyman's examination will soon be passed. A new life with more money is waving soon. And that brings advantages.

But after a year there is another obstacle. The guild of North Carolina artisans has decided: all artisans should be better educated. From now on, all plumbers in the state of North Carolina need eight years of elementary school. If you don't have that, you can catch up on the eighth school year in evening courses. Our professor alias plumber journeyman must now go to school three times a week in the evening.

The first lesson is on mathematics. The teacher enters.
*with emphasis*
"Hello guys. I want to see what you know so far. Tell me, what is the area of a circle."

Silence. "It is round."

The teacher distorts his face. "But how big is it?" An awkward silence again. Finally our professor lifts his hand. He has forgotten, but can derive it. Whether he should come to the blackboard. Yes, so he goes to the board and starts to calculate:[1] *slowly, like a professor*

Integral from r to minus r over the root of r-square minus x-square de x

is equal to integral from r to minus r fraction r-square minus x-square over the root of r-square minus x-square de x

is equal to and so on...

---

[1] $\int_r^{-r} \sqrt{r^2 - x^2}\, dx = \int_r^{-r} \frac{r^2 - x^2}{\sqrt{r^2 - x^2}}\, dx = \ldots$

The professor continues to lecture: "The next step is to separate the integral into a sum of two integrals, and then one needs an integration by parts."

The formulas become longer and longer, the teacher becomes anxious and more anxious. Some of the schoolboys have already fallen half asleep, others grin with malicious glee. But the professor does his job and at the end of the calculation he gets "minus pi times r-squared half." $(-\pi r^2/2)$ But that's not what one expects. The professor sweats, and even the teacher sweats a little bit. Finally, one agrees that this is only the area above the x-axis, and you have to add the same area below the x-axis. So you get the double, which is "minus pi times r-squared." But that's not as expected, either. The professor sweats even worse, and the teacher almost gives up. Finally, the smallest guy answers from the last row.

"One should swap the boundaries of the integral, and calculate integral from minus r to r."

*angry bass*

Bad tongues claim that the professor got a D and the clever brush an A in math. And so the professor is still a plumber today. But the guy, yes, that's a different story.

**Remark**. The professor has answered the question of the teacher as part of a top quality calculus course. But calculus, as part of the mathematics curriculum, is not based on a completed axiomatic system. Such a course usually does not give an analytical definition of trigonometric functions, nor of the number $\pi$. These notions are usually taken as assumed knowledge from school geometry. The professor's explanations are not self-contained. In a calculus class, the professor could live with these caviets.

The teacher, too, does not try to provide self-contained explanations. All that should be clear to the professor. But he is already confused by having to participate in that class at school. He cannot imagine himself back to school, sitting in such a class. This explains his ridiculous reaction.

How could he have responded better? Just tell the class: the area of a circle equals $\pi$ times the radius squared. Leave it to the teacher to pose further questions.

## 6.2 Musil's Törleß on the Theater

A theater adaptation of the "Young Törleß" was to be seen on stage in Mainz, with a woman as director and with women in the leading roles. Even the trailer shows the striking style aimed at excess, everything stylized and abstract to the extreme. On the contrary, the mastery of the novel lies in describing the dangerously slow way in which the young pupils ("Zöglinge") are drawn step by step into the whirlwind of masochism and sadism, medieval hypnosis, and more strayings. Smart actors are needed to slowly increase the tension. They act in and as the center, they are the most important part of the performance. Too much artistry, distorting speakers, and more of the like, all that only distracts the audience. I have already observed similar defects when comparing two versions of Dürrenmatt's physicists: broadcast on television in 1964 on the one hand, and with Herbert Fritsch in 2015 on the other hand. In this case, too, one can see how the artistry takes over, thus pushes back, and meddles with the content of the piece. I now understand why Dürrenmatt turned away from the Zurich theater, becoming rather disappointed towards the end of his career. After a long talk, the reasons are becoming apparent during the discussions with Ludwig Arnold.

It takes a lot of skills to show both strands of the Törleß novel: on the one hand, the confusion that mathematics creates in Törleß as soon as you think beyond the small multiplication table, —on the other hand, the extremely strong relationship between Törleß and his mother. This unhealthy bond also favors perversions-sadism, masochism, even including homosexuality. Puberty and growing up are for the youngsters always cause both cognitive and emotional problems. Stressing this double challenge is for me the essence of the novel *"The Confusions of the Pupil Törleß "*.

It becomes obvious how dangerous both incomplete knowledge and bragging are, especially during puberty,—how they favor latent mental disorders and cause them to break through. Can you convert "The Young Törleß" into a good play that shows the above processes even more emphatically than Schlöndorff's movie? It would need a master to do this.

My goal is to expose more clearly what is confusing the young Törleß, on the one side within mathematics, on the other side about the life of the adults in general. Here are a few lines as a first humble beginning of what I intent.

> Teacher: Today, I want to explain the imaginary numbers. The idea is very simple. We just pretend there is something like "the root of minus one", and we use this object as a unit of account.
>
> Törleß: Teacher, but there is no root of minus one at all.
>
> Beineberg: It doesn't matter, we just pretend it exists.
>
> Teacher: Beineberg always participates well.
>
> Törleß: Sir, may I write my formula onto the blackboard, it confuses me. Everything is correct, but why is it not working?
>
> Törleß writes on the board: [2]
> *slowly, like a professor*
>
>> minus one equals the root of minus one squared is equal to the root of minus one times the root of minus one
>>
>> second line: is equal to the root of in brackets minus one times minus one is equal to the root of one equals one.
>
> Törleß: In the end we get that minus one equals one. But we know that this is wrong.

---

[2] $$-1 = (\sqrt{-1})^2 = \sqrt{-1} \cdot \sqrt{-1}$$
$$= \sqrt{(-1)(-1)} = \sqrt{1} = 1$$

Teacher: Later you will understand all of this. Now it's just a matter of believing that it works.

Reiting: Caesar says: the first line is allowed because I saw it, allowed it, and won. The second is forbidden because it brings us to the deepest Germania. It must never be written down again.

Beineberg: No, I have to see what's on the board. How the roots touch, loop around one another, and unite. I have to go right through this struggle. That is my way.

Basini: If someone simply takes a euro away, it won't be right, either.

Next appearance. Already on their way, they talk to each other.

Reiting: I am lacking money and Basini reveals himself. Törleß: But he's so honest.

Reiting: I really think Basini took a euro out of my drawer. He wants to distract us. Otherwise, he wouldn't have talked about taking away a euro.

Beineberg: The interrogation in the little closet will be our pleasure.

Törleß: What do you find out. You won't be able to calculate and build bridges.

Bozena: Look, how cute boys are coming! But why so excited, did the teacher confuse you once more?

Beineberg: It is not his profession to show how the roots unite.

Bozena: With me, you are at the source. Not only roots unite here.

Basini: My whole family comes from noble roots. When we all gather, we fill a large hall.

Beineberg: I just want you. The unification needs practice and concentration.

Bozena: For the beginning, the boys may get some tea and rum. Well, fresh up!

**Remark**. The answers of Reiting and of Beineberg both refer to mathematically correct facts, whereas the teacher understands almost nothing.

Indeed, as long as one restricts oneself to *basic arithmetic*, that is addition, subtraction, multiplication, and division, the new rule $i^2 = -1$ correctly leads to the field of complex numbers.

But the calculation with roots is not possible within this field. All that one can achieve is a restriction to multiplication, division and taking square roots, leaving apart addition and sub traction. To this end, one needs to construct the *Riemannian surface* of the square root. This surface winds itself twice around the zero point, before closing itself.

So far about some kind of bad dreams that sometimes overwhelm me. In reality, Franz Rothe graduated from high school in Karlsruhe and has studied mathematics, physics and music there. He has received his doctoral degree in mathematics from the university of Tübingen, Germany. He was professor at the University of North Carolina at Charlotte, and has published about 40 articles and a lecture notes in mathematics, and more recently further books on number theory, modern algebra, graph theory and geometry. Dr. Rothe is retired since several years, and is now emeritus professor.

## 6.3    Remarks on my Work *A Course in Old and New Geometry*

Let me start that section with a nice song, a present by Alvi White.

Euclidean Skies A Dance with Angles

Geometry is, like all mathematics, built on these three main pillars: logical development, problem solving, and computation. Logical development has been perfected, mainly by Euclid and by Hilbert, to the *axiomatic method*.

But without interesting and challenging problems, this method could not give enough real life to mathematics. Many interesting construction problem are an essential part of geometry. Only challenging problems bring a hands-on more concrete character, and make the main part of every good curriculum. In my present work, both the construction problems as well as the proofs are very important. Both are illustrated with numerous color coded figures.

The entire work is based on the *"axiomatic method"*. This notion goes back to Hilbert to describe the first part of his formalist program. The essentials were already developed in ancient times in Euclid's Elements, and refined in Hilbert's foundations of geometry of 1899.

Euclid's Elements is the oldest surviving work in which mathematical subjects were developed from scratch in a thorough, rigorous and axiomatic way. He puts his principles at the beginning of the Elements and names them common notions and postulates. In place of the common notions, today are put the **logical axioms**. In place of the postulates, one has the **proper axioms**, which are specific to the (mostly first-order) language and the subject under consideration. Classically, postulates were supposed to be evident truths. Euclid, too, still considers his axioms as self-evident truths. The independence of the parallel axiom as well as the discovery of an entire bunch of different algebraic and topological structures has shattered that attitude since the nineteenth century. Instead, in modern mathematics the axioms appear in different groups depending on the subject area in question. Indeed they have become the main ingredient *defining* the different topics of mathematics. Nevertheless, even today, to build any meaningful mathematical theory, the formulas taken to be axioms should be as few as possible, and they have to be based on as simple as possible principles. They should be justified on intrinsic and extrinsic reasons, in other words ***well motivated and needed for the consequential development.***

The *formalist program* goes beyond the classical axiomatic approach from antiquity still in another important feature. There are explicitly defined not only the language and axioms to be used, but also the rules of inference. In the usual approach to first-order logic, two features of the rules of inference are worth noting:

- first, they are based entirely on logic;
- second, they are the <u>only</u> way of generating new steps and and proving any theorem.

*Without rules of inference no mathematics can be done, — any axiomatic system would be useless. Secondly, from the formalist viewpoint axioms are <u>no longer</u> considered to be absolute truths.* Instead they need to be *well motivated and a good starting point for the consequential development of a theory.* As a consequence, geometry has developed into an entire bouquet of different systems. That is often indicated by the use of the plural "geometries".

Recently, I have observed that my efforts are not the end of the surprising developments in geometry. In his thesis *On the Formalization of Foundations of Geometry*, Pierre Boutry from the group of Julien Narboux at the university of Strasbourg in France, is going further with the formalization of geometry by means of automatic reasoning.

# A Course in Old and New Geometry

## Volume I: Axiomatic and Neutral Geometry

### *Franz Rothe*

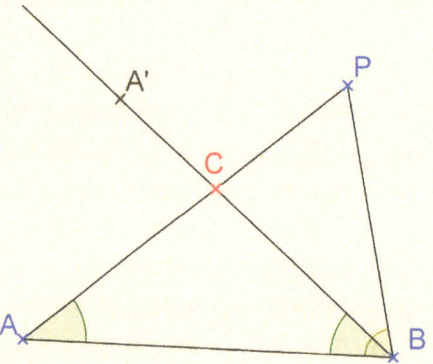

# A Course in Old and New Geometry

## Volume II:  Basic Euclidean Geometry

### *Franz Rothe*

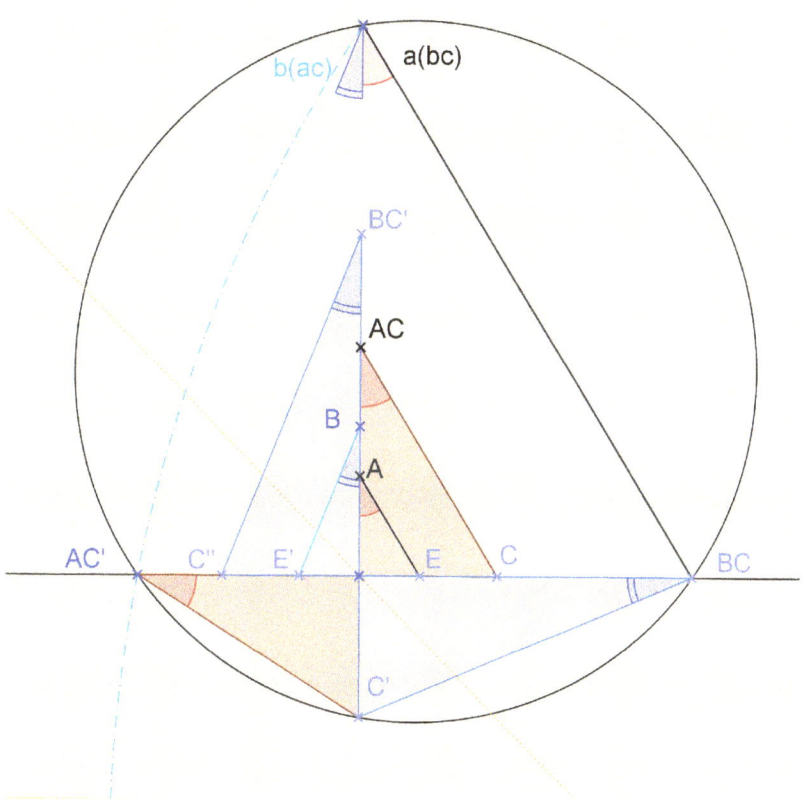

# A Course in Old and New Geometry

## Volume III:  Advanced Neutral and
## Euclidean Geometry

*Franz Rothe*

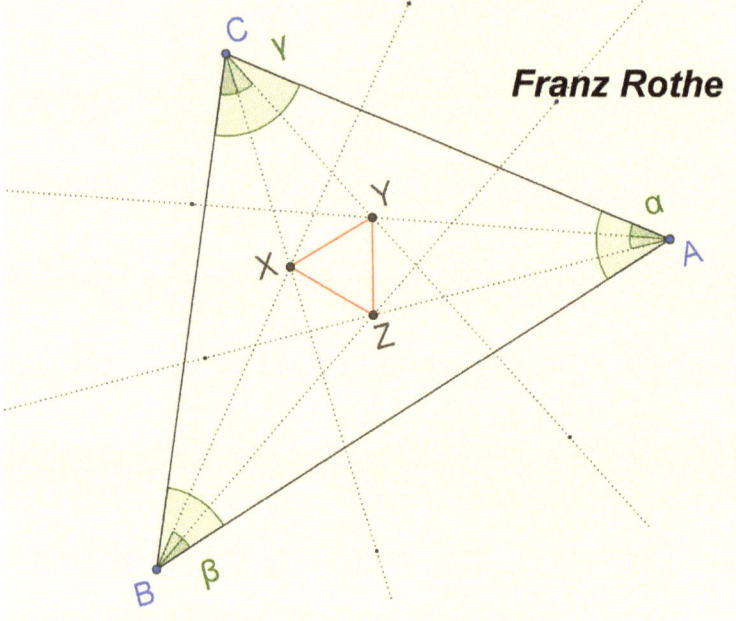

# A Course in Old and New Geometry

## Volume IV: Theory of Euclidean Constructions

### *Franz Rothe*

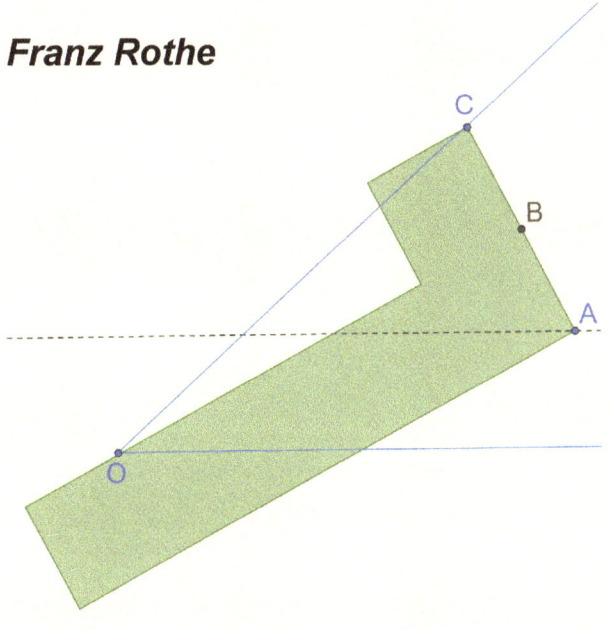

# A Course in Old and New Geometry

## Volume V:  Hyperbolic Geometry

### *Franz Rothe*

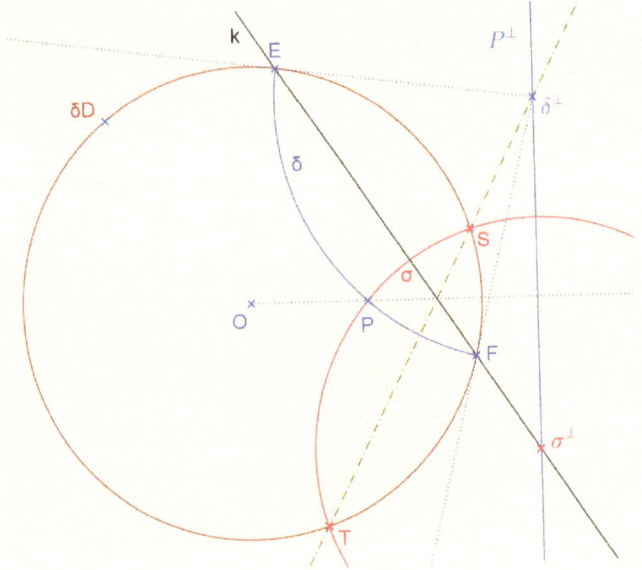

# A Course in Old and New Geometry

## Volume VI: Projective and Finite Geometry

### *Franz Rothe*

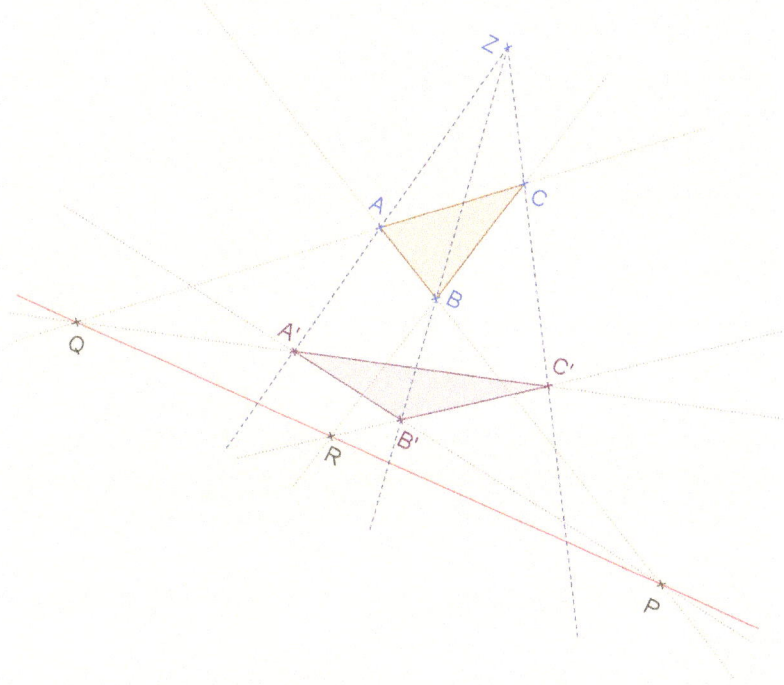

My six volume work *A Course in Old and New Geometry* is now available online from Barnes and Noble, like all my books are.

Vol I   Axiomatic and Neutral Geometry

Vol II  Basic Euclidean Geometry

Vol III Advanced Neutral and Euclidean Geometry

Vol IV Theory of Euclidean Constructions

Vol V   Hyperbolic Geometry

Vol VI Projective and Finite Geometry

A decade long experience of teaching the course *Fundamental of Geometry* at the University of North Carolina at Charlotte, many notes for exercises, thorough reading and research are the bases for this bulky work. My heartfelt thanks go to everyone involved, this means famous authors as well as my former students. Retired I could find the time and leisure to prepare these six volumes of "A Course in Old and New Geometry". Each one of the six volumes contains a review of enough material from the other volumes to be self contained. The goal of the subdivision is to obtain self-contained books concentrating on one or two topics and easy to handle.

The first volume with the subtitle *Axiomatic and Neutral Geometry* begins with Hilbert's axioms from his book *Foundations of Geometry*. After some discussion of logic and axioms in general, we go on with a short section about incidence geometries in two and three dimensions.

As in Hilbert's system, there follow detailed sections about the axioms of order, next about congruence in neutral geometry,—which means without assumption of the axiom of parallelism. I deal with the axioms of measurement and of completeness, and deviating from Hilbert, with axioms about circles. The last two sections deal with the theory of area in neutral geometry, and neutral triangle geometry.

The second volume with the subtitle *Basic Euclidean Geometry* starts with Euclidean geometry, and the text gets a more educational and even more elementary flavour. A common beginning is Thales' theorem about the angle in a semicircle, and I continue with Euclid's related theorems about angles in a circle. The most simple parts of the Euclidean geometry are given in detail, as well as the later parts about similarity, and finally area in Euclidean geometry with the theorem of Pythagoras and trigonometry, and the measurement of the circle. Too, the lens equation from geometrical optics is treated.

In several sections is recalled the strictly modern view of the first volume, beginning with Hilbert's axioms from the *Foundations of Geometry*. The theorems from neutral geometry which have been proved in the first volume and are needed again, are cited. The relation of analytic and synthetic geometry is treated on Hilbert's rigorous account.

The third volume with the subtitle *Advanced Neutral and Euclidean Geometry* elaborates many faces of advanced neutral as well as Euclidean geometry. A first part about neutral geometry originates from the century long efforts to prove the parallel axiom, from Proclus in antiquity to Legendre and his contemporaries. The insight of independence of the parallel axiom leaves many open roads to pursue, indeed the desire to develop a natural as well as completely axiomatic system still leads to new developments. In this context, the classification of Hilbert planes into three types, —as semi-euclidean, semi-elliptic or semi-hyperbolic, known as the *Uniformity Theorem* is an important step beyond Euclid. A further step is the introduction of Aristole's axiom about the unbounded opening of an angle, which has been suggested by Greenberg in his book *Euclidean and Non-Euclidean Geometry*, and leads to a natural classification of geometries. The further part about advanced Euclidean geometry includes a thorough triangle geometry, harmonic points, some elliptic geometry, and many more interesting problems.

The fourth volume with the subtitle *Theory of Euclidean Constructions* elaborates the theory of geometric constructions.

Constructions with both extensions and restrictions of the classical Euclidean tools, straightedge and compass, are explained and investigated. One finds constructions by a marked ruler, which allows the duplication of the cube and angle trisection and the construction of a regular seven-gon. On the other hand proofs for the impossibility of such constructions by the Euclidean tools or Hilbert tools are obtained with tools of modern algebra. The connection to modern algebra, and especially Galois theory is explained. Detailed attention is given to Alhazen's problem about reflection by a circular mirror, the lunes of Hippocrates and their relation to transcendental numbers and, the quadratrix of Hippias.

The fifth volume with the subtitle *Hyperbolic Geometry* recalls Hilbert's axioms from the *Foundations of Geometry*, and elaborates hyperbolic geometry. Here the disk models of Poincaré and Klein are used to do a lot of constructions, using straight-edge and compass from the background Euclidean geometry. Too, Hilbert's axiomatic approach based on the asymptotic rays, is explained from the beginning up to the reconstruction of the Poincaré disk model. The last section gives a short course on Gauss' differential geometry and the pseudo sphere.

The sixth volume with the subtitle *Projective and Finite Geometry* reviews Hilbert's axioms from the *Foundations of Geometry*. After some discussion of logic and axioms in general, we go on with a thorough treatment about incidence geometries in two and three dimensions. Here the Theorems of Desargues and Pappus and their relatives, and their relation to geometry built from coordinates are discussed. Hilbert points to these development already in the introduction his *Foundations of Geometry*. With his own words: "We shall be challenged by very new and—as I believe fruitful—problems, and see remarkable connections between the elements of arithmetic and geometry, gaining another insight into the unity of mathematics."

In further sections, the perspective view, and Pascal's contributions are explained. The final section concentrates on finite geometries which have close connections to large scale modern computations.

The entire work is based on Hilbert's *Foundations of Geometry*. The covered topics and the presentation have profited much from Marvin J. Greenberg, *Euclidean and Non-Euclidean Geometry*, and Robin Hartshorne, *Geometry: Euclid and Beyond*. But the work contains several topics which are not covered in these books. Too, numerous very detailed proofs, solutions of problems, and the illustrations with numerous color coded figures, is new.

## 6.4    Why Axiomatic Set Theory?

Intuitively, a set is a collection of all elements that satisfy a certain given property. Cantor gives the following, rather naïve definition for a set:

> "A set is a collection into a whole of definite distinct objects of our intuition or of our thought. The objects are called the elements (members) of the set".

The above citation is following Abraham A. Fraenkel's book *Foundations of Set Theory* . Cantor's sentence sounds so harmlessly nice and easy to understand. It leads to the paradise-like infinities too quickly. With or without paradoxes it soon becomes obvious that well-defined terms are needed.

So it is better to speak from the very beginning about a *set* and its *elements*, and to introduce the relation "Element lies in the set" at the beginning—written as a formula:

> "x is the element of A." or simply $x \in A$.

I didn't find this relation— "is an element of"—in Cantor's view of set theory. It probably goes back to Bolzano, or to Zermelo. Gradually, it may also dawn on the reader that *well-defined terms* are requiring more than simply generally understandable language. Additionally, these terms should and must be clearly exemplified. The first steps in the foundations of mathematics, i.e. the basic concepts in logic, set theory, and algorithms form the first part of Hilbert's program, and these are common knowledge for the modern mathematicians of today. I say and write: *these basic concepts have become commonplace.* Therefore, the first part of Hilbert's program is a major step forward. The mathematician Nelson, in a recent review of the history of foundations of mathematics, even speaks of a further progressive step of the enlightenment.

### 6.4.1 The Naïve Principle of Comprehension

Although elementary set theory fills Cantor's naïve definition of sets at least with a bid of substance, the central point of critic remains: naïve set theory is based on the assumption that a general comprehension principle is valid. In other words, we might be tempted to postulate the following principle:

**(Naïve principle of comprehension).** *For any property P (x) on x there exists a set which contains exactly those elements x which fulfil this property.*

Here is Russell's famous reasoning showing that this principle cannot be claimed to be true. Such a general principle would stipulate that for any well-formed formula $\varphi$ with the free variable $x$, the propositional function $\varphi(x)$ one can define the set $A$ containing all elements $x$ with the property $\varphi(x)$

[3]For this set would hold

$x$ is an element of $A$ if and only if the property $\varphi(x)$ holds.

---

[3]     written as the formula $A = \{x : \varphi(x)\}$

[4]Indeed, Russell takes for $\varphi(x)$ the formula *"x is not contained in x"*, written as the formula $x \notin x$. According to the explanations below, this is indeed a well-formed formula. The general comprehension principle would allow us to define a set $A$ such that the equivalence

> *"x is an element of A if and only if x is not an element of x"*

holds for all $x$. Substituting $A$ for the variable $x$ leads to the formula

> *"A is an element of A if and only if A is not an element of A",*

which is a clear contradiction!

**Theorem 2.** *The <u>general comprehension principle is not valid</u>. Especially we have a counterexample:*

> *There does not exist a set y such that for all x holds:*
> *x is an element of y if and only if x is not element of x*

[5]Russell's paradox is nothing else than a proof that the general comprehension principle is <u>not</u> valid. Because of the strong intuitive appeal of the general comprehension principle, this proof is called a "paradox".

**Remark.** Together with Hans-Peter, we are still speculating about Georg Cantor, the founder of set theory. Like some of his colleagues from the humanities, H.P. believes in a case of spiritual self-abandonment. Let me in the following make a few remarks about this matter.

Certainly there is a hiatus of many years in Cantor's life, during which he did not and probably could not deal with

---

[4]　　written as the formula $\forall x\, [x \in A \Longleftrightarrow \varphi(x)]$

[5]　　written as the formula $\neg \exists y \forall x (x \in y \Longleftrightarrow x \notin x)$

mathematics. In Cantor's letters from these years, one reads that he has not felt so fresh lately, and among other topics, one finds speculation about the origin of Shakespeare's dramas. In any case, after several years he straightens up again, and published a second proof for the uncountability of the real numbers, and presented this piece of research at the founding meeting of the DMV (German Mathematics Association).

This famous *diagonal argument* is closely related to Russell's paradox. A little later he was familiar with the paradoxes of set theory. According to some biographers, they didn't throw him off track. These biographers say that he would have had psychological problems, too, if he had been simply a baker.

### 6.4.2  About Further "Paradoxes"

However, since this general comprehension principle is so close to our intuitive concept of set we shall try to retain a considerable number of instances of this axiom schema. But this is not as easy as it might seem at first. The instance which we used here to get a contradiction is by no means the only contradictory instance of the axiom schema of comprehension; moreover, there are non-contradictory instances of this axiom schema which contradict each other. A second serious problem with the naïve principle of comprehension arises from the semantic paradoxes. A typical one is Berry's paradox:

> For the sake of argument, let us admit that all the words of the English language are listed in some standard dictionary. Let $T$ be the set of all the natural numbers that "can be described in fewer than twenty words of the English language". Since there are only a finite number of English words, there are only finitely many combinations of fewer than twenty such words, hence $T$ is a finite set. Quite obviously, there are natural numbers which are greater than all the elements of $T$. Hence there exists a least natural number which

cannot be described in fewer than twenty words of the English language. And by its definition, this number is not in the set $T$.

But wait a moment: yet we have described this very number in sixteen words, hence it is in $T$. Once again, we are faced with a glaring contradiction.

Once one accepts the English language, the above argument would be unimpeachable. If we admitted the existence of the set $T$, we are irrevocably led to the contradicting conclusion that a set such as $T$ simply cannot exist.

How can the contradiction be resolved? The current answer is: one has to define precisely what a "language" is. The type of restricted language acceptable in mathematical logics is called a *"first order-language"*. Such a language $\mathcal{L}$ is defined by a stepwise construction. This procedure is called a *recursive definition*. One begins with *variables*, of which have to be infinity many instances. The construction proceeds to the *well-formed formulas*. For example $x = x$, $x \neq x$, $x \notin x$ or $x \in x$ are such well-formed formulas. Following von Neumann, the *natural numbers* are define by setting $0 = \varnothing$, and inductively $n = \{0, \ldots, n-1\}$ for all natural numbers $n$. If the English language is replaced by such a first-order language, in place of the set $T$, one obtains a legimitely constructed set $T'$.

Now one is able to correctly state: there exists

> *"a least natural number n which cannot be described in fewer than twenty words of the language $\mathcal{L}$"*.

Now we have described this number with 18 words. It should be in the set $T'$. Above is stated by definition the contrary, a contradiction!

Once more, how is the contradiction avoided? The answer is: The number $n$ is correctly constructed using the *everyday language*, But it is not immediately clear how to translate the above sentence into an explanation of the construction *"within the first-order language $\mathcal{L}$"*. I have now two cases to consider:

(a) The translation of the above construction into one within the language $L$ does not exist.

(b) The translation of the above construction into one within the language $L$ exists.

In case (a) it is not possible to argue as about the length of the translation. No contradiction arises. In case (b) there exist such a translation. Hence it has some,—possibly very large but finite length. A contradiction would arise only in the case that this translation is less than twenty words long. The above argument is turned into a proof that each such translation is at least twenty words long. No short enough translation exists.

I am not clever enough to prove whether case (a) or case (b) occurs, although I firmly believe case (b) is true. But that does not matter, to my great luck!

### 6.4.3   Zermelo's Set Theory

Before the paradoxes, the question of the existence of sets had never been posed. Now it is evident that the axioms are needed to assure which sets still exist.

Zermelo was the first to create, around 1908, a set of axioms for set theory. Zermelo took a pragmatic view of the problem. No doubt the general comprehension principle was not valid, Russell's paradox had made that clear. On the other hand, looking at the way the principle is used to justify the basic facts about sets, there are used only a few, simple and seemingly noncontradictory specific applications of this principle. Based on these lines of thought, Zermelo replaced the unrestricted (and indeed false!) comprehension principle with a more cautious list of principles for the existence of sets. Zermelo writes

> Under such circumstances there is at this point nothing left for us to do but to proceed in the opposite direction, [away from the general principle of comprehension] and, starting from set theory as it is historically given,

to seek out the principles required for establishing the foundations of this mathematical discipline. In solving the problem, we must, on the one hand, restrict these principles sufficiently to exclude all contradictions and, on the other hand, take them sufficiently wide to retain all that is valuable in this theory.

In other words, Zermelo proposed to replace the direct intuitions about sets, as Cantor has taken them for granted, with some axioms. These are hypotheses we accept, with little a priori justification, simply because they are necessary for the proofs of the fundamental results of the existing theory, and seemingly free of contradictions.

Within Zermelo's axiomatic set theory, Russell's paradox does not exist any longer. Instead, repeating the same reasoning leads to the proposition explained and proved below.

### 6.4.4　Answer to Russell

The following is an important axiom in Zermelo's system of axioms for set theory. [6]It is called *axiom of separation* since from any set $A$ and any property $P(x)$, one obtains the subset of elements satisfying that property, and the complementary subset not satisfying that property. Too, as already mentioned above, one needs a first-order language to define what is a legitimate property.

(**Axiom of separation**). For any set $A$ and any property $P(x)$ there exists the set of all elements from $A$ which have the property $P(x)$

Calling this set for example $S$, one may define a formula

$$S := \{x \in A : x \text{ has the property } P(x)\}$$

---

[6]　The remaining axioms are very few, and even easier to understand. It goes beyond the scope of this small book to go further into these matters.

Hence for this set $S$ we know that

$$x \in S \text{ if and only if } x \in A \text{ and property } P(x) \text{ holds}$$

**Problem 1.** *Given any set A we define the set R containing all elements x from the set A which satisfy the formula $x \notin x$.*[7]

(a)  *For which elements holds $x \in R$.*

(b)  *Convince yourself that $A \notin R$.*

(c)  *Convince yourself that $R \notin R$.*

(d)  *Give a reason why $R \notin A$.*

(e)  *In which way can one think positively about the two latter results?*

Answer. (a)

$$x \in R \text{ if and only if } x \in A \text{ and } x \notin x$$

(b) Substitute $A$ for $x$. You see that $A \in R$ implies both $A \in A$ and $A \notin A$, which is a contradiction. Hence $A \notin R$.

(c) Substitute $R$ for $x$. You see that $R \in R$ implies both $R \in A$ and $R \notin R$, which is a contradiction. Hence holds $R \notin R$.

(d) The negation of item (a) is

$$x \notin R \text{ if and only if } x \notin A \text{ or } x \in x$$

Substituting $R$ for $x$ tells

$$R \notin R \text{ if and only if } R \notin A \text{ or } R \in R$$

Since $R \notin R$, the rule of detachment yields $R \notin A$ or $R \in R$. Since the second alternative was ruled out, we get $R \notin A$, as claimed.

---

[7]  Written as the formula $R := \{x \in A : x \notin / x\}$

(e) For any set $A$, we can find a variable $R$ not contained in the set $A$. The same set $R$ does not contain $A$.

**Remark.** The above explanations clarify two important possible paradoxes, going back to Russell and to Berry. But is now once and for all excluded the possibility of further paradoxes? The answer is no. The axioms are like a fence around mathematics, but one cannot prove that no wolves (contradictions) are already inside the fence. The mathematicians living inside the fence, —now easily mixed up with sheep, —have to deal with the danger around them. This nice aphorism goes back to Henri Poincaré.

## 6.5    My Invitation to Mathematical Problems

The topics of this book come from a variety of fields, and considerable impact is given to computation with mathematica. It is my hope to keep alive the interest of a broader group of young and older readers, who are not mainly concerned to dig themselves into a single field of speciality. This book has been written to sharpen the mathematical abilities of the reader, and my own ones, too, —while keeping an equilibrium of entertainment and challenge. The last chapter deals with some mathematical physics. It contains explicit calculations for the rainbow, and the reasoning leading to the universal Plank units.

# My Invitation to Mathematical Problems

## Computations, Solutions, Proofs and Insights

### Franz Rothe

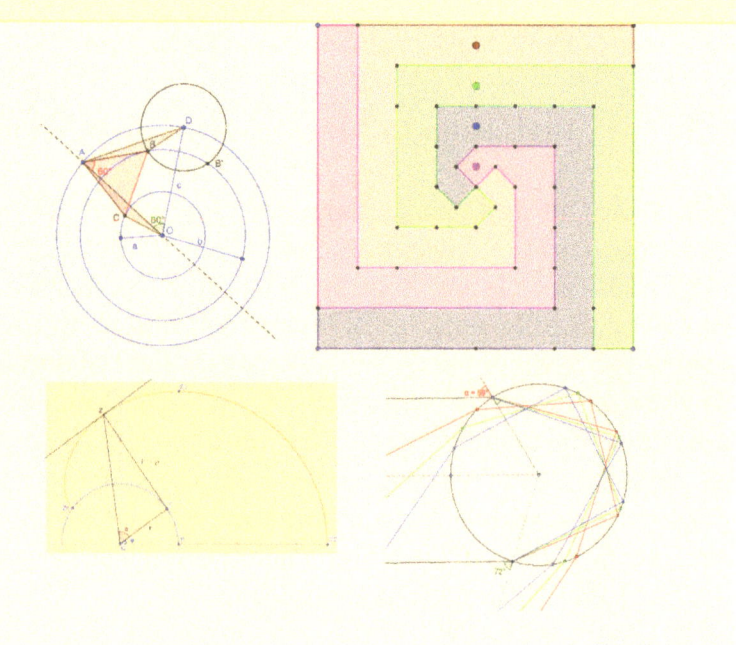

# 6.6    Essays about Modern Algebra: *From Classical Topics to New Problems*

This book has been growing over the years out of several resources. Firstly, the courses in number theory and modern algebra I had been teaching at UNC Charlotte. Secondly, my interest to invent mathematical problems for competitions, and individual work with gifted students.

Besides complete proofs, simple as well as advanced problems, finally computation on all levels are elaborated. Simple algorithms for prime factorization, the Euclidean algorithm and the Chinese remainder problem are even given for the pocket calculator. The construction of Fermat polygons is done via computations with mathematica, and a second computer language, finally numeric calculations for check of correctness. Luckily enough, with these means, I could completely work out the constructions of the regular 17, 257 and even the 65 537-gon. This material is much more concrete and computational than a course in modern algebra usually is.

Complete proof on quadratic reciprocity is given. I prove the basic theorems on totally real and totally positive algebraic numbers. These numbers come up in the geometric construction by Hilbert tools, —which are a bid more restrictive tools than the classical tools compass and straightedge. The theory of geometric constructions with different sets of tools, —starting from Hilbert tools, via the classical compass and straightedge, and finally the use of two-marked straightedge and their treatment with Galois theory, —are explained in all details. The Galois theory is treated in Artin's elegant approach via characters. Thus both characteristic zero and prime characteristic are covered.

Many generally known as well own problems about prime numbers, the Euler totient function, the cyclotomic and other special polynomials, formulas for $\pi$, are solved completely. In the end, the shear size of material I had gathered made it convenient and necessary to take a break, and think about publication.

Here is a paragraph by Hermann Weyl fitting nicely to the intention of the present book.

> Important though the general concepts and propositions may be with which the modern and industrious passion for axiomatizing and generalizing has presented us, in algebra perhaps more than everywhere else, nevertheless I am convinced that the special problems in all their complexity constitute the stock and core of mathematics, and that to master their difficulties requires on the whole the harder labor.

# Essays about Modern Algebra

## From Classical Topics to New Problems

## *Franz Rothe*

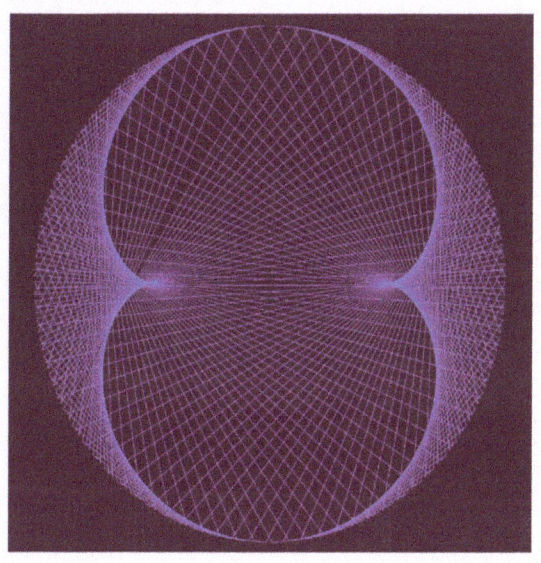

## 6.7   A Closing Word

Meanwhile, I have two cats. The small male tiger likes to go for a walk during nicer days, and hopefully always will come back onto my lap. The black and white big female one accompanies me during my lonely nights.

The angels are able to provide plenty of joy and relief, and occasional help, but they cannot free anybody from basic serious obligations, —they cannot do the work of men nor correct missed duties. I guess: in some sense they are similar to the cats. I close with a word from old Goethe, which I have translated as good as it is possible for me:

> Young man, remember in times,
> When mind and spirits rise:
> Accompanying the muse can,
> But conducting does not understand.

> —*Johann Wolfgang von Goethe*

How did the decision to write such a book come about?

A smart schoolmate invited each of our final class members to write about themselves. Using some short essays I enjoyed writing, I had gathered enough material for such a booklet. The task of creating a comprehensive autobiography seemed too changing. So I decided to gather my short observations, put them into a loose context and make a book, even though it is clearly incomplete.

Why do you think this book will appeal to more people than only your closest friends?

The description of professional growth, and the music playing which I have practiced with endurance during my entire life can motivate and benefit a broader audience.

 *Franz Rothe* graduated from high school in Karlsruhe and has studied mathematics, physics and music there. He has received his doctoral degree in mathematics from the university of Tübingen, Germany. He received his doctoral degree in mathematics from the university of Tübingen, Germany. He got Habilitation and venia legendi from the university at Tübingen (1984), and the Ludwig Maximilian university of Munich (1988).

For thirty years, he has been professor at the University of North Carolina at Charlotte, and has published about 40 articles and a lecture notes in mathematics, and more recently further books on number theory, modern algebra, graph theory and geometry. Dr. Rothe is retired since several years, and is now emeritus professor. Meanwhile he has written a book about music, literature, and his own life of general interest, too.